how to rebuild your SMALL-BLOCK FORD

by Tom Monroe
Registered Professional Engineer;
Member, Society of Automotive Engineers

INTRODUCTION	
Chapter 1 DO YOU NEED TO REBUILD?	3
Chapter 2 ENGINE REMOVAL	11
Chapter 3 PARTS IDENTIFICATION & INTERCHANGE	24
Chapter 4 TEARDOWN	39
Chapter 5 INSPECTING & RECONDITIONING THE SHORTBLOCK	51
Chapter 6 HEAD RECONDITIONING & ASSEMBLY	75
Chapter 7 ENGINE ASSEMBLY	93
Chapter 8 DISTRIBUTOR REBUILD	122
Chapter 9 ENGINE INSTALLATION	134
Chapter 10 TUNEUP	155
INDEX	158

ANOTHER FACT-FILLED AUTOMOTIVE BOOK FROM H.P. BOOKS

Notice: The information contained in this book is true and complete to the best of our knowledge. All recommendations on parts and procedures are made without any guarantees on the part of the author or HPBooks. Because the quality of parts, materials and methods are beyond our control, author and publisher disclaim all liability incurred in connection with the use of this information.

The cooperation of Ford Motor Company is gratefully acknowledged. However, this publication is a wholly independent production of H.P. Books (Fisher Publishing Inc.).

Publisher and Editor: Bill Fisher; **Editor-in-Chief:** Carl Shipman; **Art Director:** Don Burton; **Book Design:** Tom Jakeway: **Typography:** Cindy Coatsworth, Connie Brown, Kris Spitler; **Drawings and photos:** Tom Monroe; **Cover photo:** Don Winston.

Published by **H.P. Books**, P. O. Box 5367, Tucson, AZ 85703 602/888-2150
ISBN 0-912656-89-1
Library of Congress Card Catalog Number 78-74545 ©1978 H.P. Books
Printed in U.S.A.

289 powered 1964 Fairlane and 1966 Mustang GT, both with their original engines. The Fairlane is used daily and has traveled more than 100,000 miles, and the Mustang has gone about half the distance. These vehicles illustrate the all-important point that the most important ingredient for getting the most out of your small-block Ford is proper service and maintenance.

1962 221 CID Fairlane V-8 rated at 145 HP has road-draft crankcase-ventilation tube, generator and oil-fill tube in timing-chain cover. Photo courtesy Ford.

Introduction

The *small-block Ford V8* is one of the more substantial engine families in the Ford line of V8 engines that began in 1932. The small-block's life began in 1962 as the *Fairlane V8*, because it was designed for and installed in the Fairlane car. Because the Fairlane was then a mid-sized car, the engine was small in size *and* weight.

The original displacement was 221 cubic inches—3.50-in. bore and 2.87-in. stroke. A 0.30-in. bore increase later in the same year produced 260 CID, the "big" Fairlane engine. Since then, the small-block family has expanded to include 289, HP289, 302, Boss 302 and 351W engines—W indicating Windsor, Canada, where the engine is produced. W distinguishes it from the 351C, or Cleveland—same displacement, but different engine.

Good basic design is proved by its longevity and extensive applications. Since 1962 it has been installed in passenger cars, trucks, boats, sports cars, race cars—just about anything that requires power. Beginning with a win in the '65 Indy 500, it served as the basis for an all-out twin-cam race engine. More importantly, the engine is quickly becoming Ford's "big-block of the future," considering the mileage requirements imposed by the Federal government and the impending fuel shortage. I believe the future of the small-block Ford is assured.

Let's get on with what this book is about—keeping your small-block around longer by restoring it to peak mechanical condition. Rebuilding an engine seems like an ominous undertaking to the person who's never done it. And ominous it is, particularly when you consider the number of components in an engine and the decisions that have to be made during a rebuild. However, the difficulty of rebuilding an engine is reduced in direct proportion to the information you have about your engine. The more you have, the easier the job. The less you have, the more difficult it is—to the point of being impossible.

While preparing this book I rebuilt several engines and photographed every step—determining if the engine needs rebuilding, or what it needs, removing it from the vehicle, tearing it down, inspecting it, reconditioning the parts and reassembling them into a complete engine, and finally reinstalling the engine and breaking it in. I tried to leave nothing to your imagination. This book has more than typical specifications and information on how to tear an engine down and reassemble it. It includes information to make you the expert on your engine; tells what you need to do the job. I discuss what you can and can't "get away with," to keep you from being victimized by well-meaning but inaccurate information. The information I include in words *and* pictures is a product of both my experience and knowledge and that of experts who make their living rebuilding engines, tuning them and supplying new and reconditioned parts for Fords.

A final word before getting started. Work safely and use the right tool for the specific job. Take nothing for granted—check *everything*—and be alert. A bearing cap installed backwards or in the wrong location will spell the difference between a successful engine rebuild and at least a $200 disaster. With those points in mind, let's get into your engine.

Do You Need To Rebuild? 1

1965 260 CID Ford was 39 CID larger due to a new cylinder-block casting with 0.30-in. larger bore than the 221.

Before tearing down your engine, you should determine to what extent it needs rebuilding, if at all. Your engine must have thousands of miles on it, be way down on power, getting poor fuel mileage or consuming excessive amounts of oil—otherwise you shouldn't be considering a rebuild.

Just because an engine has accumulated some miles it doesn't necessarily need rebuilding. A reasonably well maintained and operated small-block Ford can exceed the 100,000-mile mark and still provide excellent service. The way an engine is used is significant. For example, an engine in an off-road vehicle naturally inhales more dust than one in the typical family sedan. More internal wear per mile of use is the result. An engine which has been frequently over-revved will also wear excessively and may have internal damage such as broken piston rings or bent valves.

EXCESSIVE OIL CONSUMPTION

The amount of oil an engine uses between changes is the "yardstick" most people use to judge the condition of their engines—and rightly so. Oil consumption is largely determined by clearances between moving parts. As the engine wears, these clearances increase, resulting in increased oil consumption *and* oil-pressure loss.

Just what is excessive oil consumption? I think an engine using oil at the rate of one quart in less than 1,000 miles needs attention. If a quart lasts only 500 miles or less, the condition is serious.

Other than the obvious gasket or seal leak which shows up as large oil spots on your driveway or garage floor, the two major causes of oil consumption are worn or broken piston rings and worn valve guides. These let oil escape past the pistons or intake-valve stems into the combustion chamber, or past worn exhaust-valve guides into the exhaust ports. This partially burned oil is forced out the exhaust system into the atmosphere creating an ominous puff of blue smoke from the exhaust when starting or when applying power after descending a hill.

Piston Rings—Three rings per piston are used. The top two compression rings seal compression and combustion pressures. The second compression ring also keeps oil from getting to the combustion chamber. Oil control is the job of the bottom ring—the oil ring. It doesn't completely seal the cylinder from the crankcase, otherwise the compression rings and piston wouldn't receive any lubrication and would seize or wear away the cylinder bore.

If oil can get past the rings into the combustion chamber, combustion and compression pressure can blow past the rings into the crankcase, called *blowby*. Blowby in pre-1966 engines (pre-1964 in California) pushes crankcase oil and other vapors into the atmosphere through the crankcase vent. 1966 and later engines recycle blowby into the combustion chambers through the PCV (positive crankcase ventilation) system. Therefore, worn or broken rings allow oil loss past the piston rings into the combustion chambers, out the crankcase vent or into the combustion chamber via the PCV system.

Valve Guides—A worn valve guide allows oil to get past the valve stem and into the combustion chamber or exhaust port, depending on whether it is an intake or exhaust valve. Because some oil must be used for lubrication, there should always be some controlled oil loss through the valve guides.

LOSS IN PERFORMANCE?

When I refer to performance, I mean fuel economy as well as power. As an engine wears, its performance suffers. A change in fuel economy or oil consumption is easier to monitor than the power output of your engine because you can measure it. Judging power loss can be done with a chassis dynamometer, but you can't go by your own impression because the wearing-out process and the accompanying power loss is too gradual.

If your engine's performance is suffering, but oil consumption is normal, give your car a thorough tuneup. Choose a reputable tuneup shop with a chassis dynamometer. This measures engine power at the drive wheels while critical engine functions are being checked or adjusted. You can compare horsepower readings before and after the tuneup. Just make sure that the tuneup shop you take your car to can give you horsepower readings. Many don't or can't. If the tuneup cures the problem, relax and read the rest of the book for entertainment to see all the fun you missed. If it doesn't, you'll have to read further.

As cylinder-wall and piston ring wear progress, blowby downward and oil loss up past the piston increase, resulting in oil consumption and power loss. The additional groove in the piston shown is a "heat dam" used only in the first few months of 221 production. Photo courtesy Ford.

Monitoring engine power output at the rear wheels on a chassis dynamometer is a good way of determining whether your engine is "tired" and needs rebuilding.

CAUSES OF POOR PERFORMANCE

Let's review some possible causes of performance loss. The first suspect is piston-ring and cylinder-bore wear. These cause increased oil consumption. If blowby is excessive, there will be excessive loss of compression and combustion pressure.

The engine produces less power and gets worse fuel economy. If your engine is "hurting" in both the oil-consumption and performance departments, chances are the problem is with the rings and cylinder bores. If this is not the case, look further.

If your engine is down on gas mileage and power, but not bad on oil consumption, chances are the problem is in the cylinders or valve train. This assumes the carburetion and ignition systems are in good order. Problems can include a blown head gasket, burned exhaust valves, worn camshaft lobes and lifters, and carbon buildup.

Blown Head Gasket—A blown head gasket causes compression and combustion pressure to drop much in the same manner as bad rings and cylinder bores, only worse. Pressure lost past a gasket goes into the cooling system or an adjacent cylinder. If it leaks into the cooling system, only one cylinder will be affected making the loss in power, and even gas mileage, difficult to detect.

A gasket gone *between* cylinder bores, affects two cylinders and the performance loss is easily detectable. One thing is sure, if cylinder pressure is getting into your cooling system, your engine will overheat. The cooling system will be overpressurized, forcing coolant out of the radiator.

Burned Exhaust Valves—A burned exhaust valve can't seal its combustion chamber. Therefore, the cylinder with the bad valve will be down on compression and power. This occurs more often in later engines which operate at higher temperatures to reduce emissions. The valves run at higher temperatures and are more susceptible to burning.

Worn Camshaft Lobes and Lifters—Worn camshaft lobes and lifters almost always occur together. This problem does not affect the engine's oil consumption, but really reduces its performance. The more worn cam lobes, the more performance is affected.

When a cam lobe and lifter wears, the valve they operate doesn't open enough. If it is an intake valve, a smaller fuel charge enters the combustion chamber, causing reduced performance. A similar situation exists with the exhaust valve, but by a roundabout way. If its lift is reduced, all the exhaust cannot leave the combustion chamber, consequently there is less room for a new fuel charge when it's time for the intake valve to open. The result is the same—reduced performance.

Carbon Deposits—Carbon deposits are not a direct result of how many miles are on an engine or its age, but are caused by how it is used. A vehicle used mainly for open highway use won't experience carbon buildup, assuming the carburetion is near right. However, one that is used to putter around town at 30 MPH or so may develop the problem. Carbon deposits don't require that an engine be rebuilt to remedy the problem. But, because its symptoms can fool you, I'll discuss how carbon can affect an engine and how you can remedy the problem.

Carbon buildup takes up room in the combustion chamber, raising the compression ratio. As a result, *detonation* problems may develop, usually called *pinging*. This is caused by the fuel charge *exploding* from compression rather than *burning smoothly*. Higher loads are imposed on an engine by detonation and this causes serious damage ranging from deformed main-bearing caps to broken piston rings, and even broken pistons. *Preignition* may also occur when the carbon gets hot and acts like a two-cycle model-airplane-engine glow plug, igniting the fuel charge prematurely. This potentially serious problem can melt pistons, and break piston rings. Another thing carbon deposits cause is *dieseling,* or continuing to run after the ignition is shut off—sometimes the engine turns in the wrong direction.

Detonation, preignition and dieseling don't necessarily hurt an engine's performance, but the potential damage that can be done by detonation and preignition should concern you.

Carbon deposits hurt performance in two ways. Buildup around the valves reduces the flow of gases to and from the combustion chamber, thereby reducing power. And, pieces of carbon can break off an go out the exhaust harmlessly, or end up on the piston top, between the exhaust valve and exhaust-valve seat, or between the spark-plug electrodes. Carbon on top of the piston can reduce the clearance between the piston and the head. The engine can develop a knock, giving the impression a rod bearing has gone bad when it hasn't. I don't know anything this hurts except your peace of mind.

Carbon between the exhaust valve and seat prevents the valve from closing all the way, meaning that cylinder won't

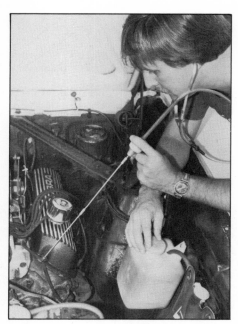

Engine problems accompanied by internal noises can be located with an automotive stethescope. You must be able to distinguish between engine noises which are normal and those which are not.

be producing much power. On the power stroke the hot fuel charge will escape between the valve and seat, overheating the valve with a good possibility of burning it.

A chunk of carbon between the spark-plug electrodes shorts the plug so it can't ignite the fuel charge. A misfiring cylinder results.

What usually causes carbon to break off and create the troubles I've just discussed is someone taking advantage of an additive sale, then going home and dumping the "instant overhaul" solution down the carburetor. The stuff *really* works, loosening the carbon which then causes these problems. Don't try the "cure-all" approach to rid your engine of carbon. Use the methods I discuss in the block and cylinder-head reconditioning chapters. Also, carbon is an effect rather than a cause. It results from an excessively rich fuel mixture, oil getting into the combustion chamber past the pistons or valve stems or very slow driving or idling for extended periods. If the carbon build-up causes are cured, the deposits gradually burn away.

DIAGNOSIS

I've discussed internal problems you may encounter with your engine and how each may affect its operation. Now let's look at how to diagnose these problems without tearing down your engine. On the other hand, you may not have any specific problems, but you do want to determine if it's time to rebuild.

Internal Noises—Perhaps your engine has noises of impending disaster coming from its innards. They may or may not be accompanied by an increase in fuel or oil consumption or a reduction in power. Generally, if the noise is at engine speed— once for every revolution of the crank— the problem is in the *bottom end*—caused by a broken piston ring, worn connecting-rod bearing or a worn pin bore in a piston.

A noise at half engine speed, or camshaft speed, is probably a valve-train problem, even though one bottom-end problem does occur at this speed. Piston slap occurs only on the power stroke, consequently it is also at half speed. If the noise is coming from the valve train it could be due to excessive *lash* or clearance in the valve train due to a collapsed hydraulic lifter, too much valve clearance or a bent pushrod. To help in determining the speed of the noise, hook up your timing light and watch the light while listening to your engine. If the light flashes in time with the noise it's at half engine speed.

To assist in listening to what's going on inside your engine use a long screwdriver and press its point against the block close to the area where you suspect the noise is coming from and the handle against your skull below your ear. This will amplify engine noises. Just make sure you place the screwdriver against a solid part of the engine to get the best noise transmittal. Don't put the end of the screwdriver against the valve cover to listen to valve-train noises, put it against a valve-cover bolt. The cork gasket between the cover and the head and the large air space under the valve cover damp out much of the noise.

What should the different noises sound like? Let's start with the engine's bottom end. A broken piston ring makes a chattering or rattling noise which is more prevalent during acceleration. A dull, or hollow sound is usually caused by piston slap or the piston wobbling and *slapping* against the side of its bore due to excess clearance between the piston and the bore. A collapsed piston skirt causes a similar, but louder noise. Slap caused by excess clearance will be loudest when the engine is cold. It decreases as the engine warms up and the piston *grows* to reduce piston-to-bore clearance. You can check for piston slap by retarding the spark. Loosen your distributor's hold-down bolt and rotate it counter-clockwise about five degrees. Retarding the spark should reduce noise due to piston slap. Use your timing light to make sure you get the distributor back in time afterwards.

A light knocking or pounding noise that's not related to detonation or pre-ignition is probably excess connecting-rod-bearing-to-journal clearance. Simply put, the bearing is worn out. Finally, a light tapping noise can indicate excess pin-bore clearance in a piston.

To confirm and pinpoint a lower-end noise-related problem, disconnect the spark plug leads one at a time, then run your engine and listen for the noise to change or go away. What happens is the power-stroke is eliminated from the cylinder with the disconnected spark-plug lead, thus unloading its connecting rod and piston. So, if the noise is piston or rod-related it will be greatly reduced or eliminated when you have the right plug wire disconnected. If your car is equipped with a solid-state ignition *always ground the lead you disconnect*. Otherwise you risk damaging your expensive ignition system.

A sharp clacking or rapping noise indicates your engine probably has a collapsed hydraulic lifter. If the noise is a light clicking, it is probably excess clearance in one of the valve mechanisms. This assumes your engine is not equipped with solid lifters which click normally— unless the clicking is excessive. Then the problem will also be excess clearance or lash. If this is the problem, a simple valve adjustment may correct the problem. To check for proper lash, remove the valve cover from the noisy side of the engine and insert a feeler gauge between each rocker arm and its valve stem one at a time. When you get to the noisy one, the gauge should take up the excess clearance and quiet the noise. In this instance a simple adjustment may correct the problem if your rocker arms are adjustable. If there is still a lot more clearance when the feeler gauge is between the rocker and the valve stem, suspect that lifter is malfunctioning or the pushrod is bent, particularly if the engine was over-revved.

Spark Plugs Tell a Story—Prior to testing, an easy way to diagnose your engine is to "look" into the combustion chambers by removing the spark plugs and inspecting them. Each plug has a story to tell about the cylinder it came out of, so remember to keep them in order.

The main thing to look for is a wet-black deposit on the inside of the threaded portion of the plug and on the plug insulator. This is caused by oil getting past worn rings or intake-valve guides. Oil loss through exhaust-valve guides won't enter the combustion chamber, and won't show on the plugs.

Black, dry and fluffy deposits are carbon caused by an over-rich fuel mixture, excessive idling or driving at sustained slow speeds without much load on the engine. All the plugs should appear about the same in this case. The fuel mixture will have to be corrected by tuning, however carbon buildup caused by the way a vehicle is operated can only be partially corrected by installing hotter plugs or by changing driving habits.

Cranking Vacuum Test—An internal-

"Reading" your spark plugs is a good way of judging your engine's condition. Worn-out spark plugs as in A are easily recognizable by eroded electrodes and pitted insulator. Replace them and your engine's performance will improve instantly. Oil-coated and fouled plug B indicates internal engine wear: piston rings, cylinder bore and valve guides. Carboned plug C coated with dry, black and fluffy deposits is usually caused by carburetion problems or driving habits. Normal spark plug D has a brown to greyish-tan appearance with some electrode wear indicated by slightly radiused electrode edges. When inspecting your plugs keep track of the cylinder each belongs to. Photos courtesy Champion Spark Plug Company.

combustion engine is a specialized air pump, so how it sucks air, or *pulls a vacuum* is an indicator of its mechanical soundness, or how your engine's cylinders are sealing relative to one another. You'll need a vacuum gauge and a remote starter or a friend to operate the starter switch while you watch the vacuum gauge. The remote starter lets you control the starter from under the hood.

When doing this test, connect the vacuum gauge to the intake manifold *after* you've warmed up your engine. Disable the ignition system so the engine can't be started. Disconnect the *high-tension*, or distributor-to-coil lead so the engine won't start when it is being cranked. With electronic systems, disconnect the distributor-to-amplifier lead. With your eye on the vacuum gauge, crank the engine. If the gauge indicates a steady vacuum reading after it has stabilized, all eight cylinders are sealing the same in relation to one another. If the needle fluctuates, indicating a pulsating vacuum, one or more of the cylinders has a problem, assuming your starter is cranking steady. It could be valve timing due to incorrect adjustment, a worn camshaft lobe or collapsed valve lifter, a leaking valve, worn cylinder bore or piston rings or a leaky head gasket. If this indicates a bad cylinder/s, the next two tests pinpoint which one it is.

Power-Balance Test—A power-balance test determines if all of your engine's cylinders are contributing the same amount of power. The method is to fast-idle your engine, then disconnect the spark plugs one-by-one and monitor the RPM drop. RPM change indicates how much each cylinder is contributing to overall engine power. The less it drops, the less it contributes. You'll need a tachometer—one that is very accurate such as found on a good *dwell tachometer*. Disconnect the spark plug leads *at the distributor* and reinstall them loosely. When removing the leads, don't pull on the wire, pull on the boot. With your engine warmed up and the tach connected, start your engine and set its idle speed to about 1,000 RPM. Remove the cylinder 1 spark-plug lead and ground it to the engine. Record engine-RPM drop after it stabilizes at the lower

SOLID-STATE IGNITION SYSTEMS
Many solid-state (electronic) ignition systems generate very high secondary voltage peaks when a spark-plug lead is unloaded by disconnecting it from its spark plug or by disconnecting the coil-to-distributor lead while the engine is being cranked or is running. The resulting secondary voltage surge (up to 60,000 volts) can damage a coil internally or pierce plug-wire insulation or a distributor cap as the surge seeks a ground. So if you remove any secondary lead with the ignition on and the engine turning, ground the lead to the engine with a jumper wire. Ford has installed solid-state (breakerless) ignitions since 1974.

Vacuum gauge being used to check the pumping ability of this engine. The needle fluctuated indicating a weak cylinder.

Two types of compression gauges were used to check the compression of this engine's cylinders. Number-8 cylinder could only muster 25 psi, whereas the others averaged 130 psi. The problem was a burnt valve. If a cylinder is down on compression, squirt some oil on the top-rear of the piston. Recheck compression a few minutes later. If compression improves, the problem is in the bore. Otherwise it's probably the valves.

level. Reconnect the lead and let the RPM return to normal, then disconnect the next lead in the firing order. When you're finished checking all of the cylinders, compare the readings. If all are within 20 RPM, no one cylinder is much worse than the others. If the drop is different by 40 RPM or more in some cylinders, then there's a problem. Note the ones with the least drop, and concentrate on them during a compression test.

Compression Testing—Comparative compression testing of an engine's cylinders gives you an idea of the condition of piston rings and cylinder bores. You'll need a compression tester, a note pad and a friend—or a remote starter switch.

Run your engine until it is up to operating temperature. Shut the engine off, disconnect the ignition wires and remove all the spark plugs, being careful not to get burned—everything is HOT. Now for the test. Prop the throttle plate open and make sure the choke plate is also open. Insert a tester in cylinder 1 spark plug hole. A screw-in type tester makes this whole procedure a lot easier. If yours has a rubber cone on the end, insert it into the spark plug hole while holding it in firmly and give it a half turn. This helps make sure it seals. Turn the engine over the same amount of strokes for each cylinder tested—about five times will do—and observe the maximum gauge pressure. Write down the results and the cylinder number. Go to cylinder 2, repeating the procedure until you've tested all cylinders in order—just to keep things organized. Now that you have the numbers, what do they mean?

You may see pressures anywhere from 80 to 250 psi, depending on the engine and its problems. The best approach is to compare the cylinder pressures with each other. All cylinders of an engine don't go bad at once, so the bad one/s will show up like a "sore thumb." Another thing to keep in mind is, cylinder pressures of a high-compression engine, such as the HP289 and Boss 302 engines, may read lower than the low-compression version of the same engine at cranking speeds. This is due to cam timing and not because of their mechanical compression ratios. This doesn't apply at higher RPM. Therefore, if you have a high-performance engine, don't be shocked if your neighbor boasts of higher cylinder pressures from his tamer regular-gas-burning engine.

What should the "spread" be between cylinder pressures? A good rule is the lowest pressure should not be less than 75 percent of the highest. Otherwise, something is not quite right in your engine. For example, if the highest reading is 150 psi and the lowest is 120 psi, then the engine is all right because 150 X 0.75 = 112—the lowest allowable pressure.

Now what do you do if all your readings are not within the 75-percent range? Something is wrong, but what? To test the piston rings, squirt about a teaspoon of heavy oil in the bad cylinder/s from your oil can—40 weight will do. The best way of determining if you are putting in the right amount of oil is to see how many squirts it takes to fill a teaspoon. Squirt the same amount of oil in the spark plug hole while directing it toward the far side of the cylinder. It will also help if you have the piston part way down the cylinder to ensure getting the oil all the way around the piston. Don't get any oil on the valves. The oil will take a little time to run down around the rings, so wait a couple of minutes before rechecking pressures. Now, if this causes the questionable cylinders to increase in pressure, the rings and bore are at fault and a rebuild is in order. If the pressure doesn't change an appreciable amount, look further.

Leak-Down Testing—Leak-down testing is similar to compression testing just covered, but rather than the engine doing the compressing, a *leak-down tester* does it. It applies a *test pressure* via a special fitting through the spark plug hole to the cylinder being tested and monitors the pressure the cylinder can maintain in relation to the test pressure. This is a better way to test an engine's condition. It eliminates factors which affect the results of a conventional compression test, but do not reflect the sealing qualities of a cylinder: valve timing, camshaft wear or engine cranking speed. The problem with a leak-down tester is it costs about three times as much as the conventional compression tester. Consequently, it's not practical for you to purchase such an expensive piece of equipment for a "one-shot deal." Leak-down

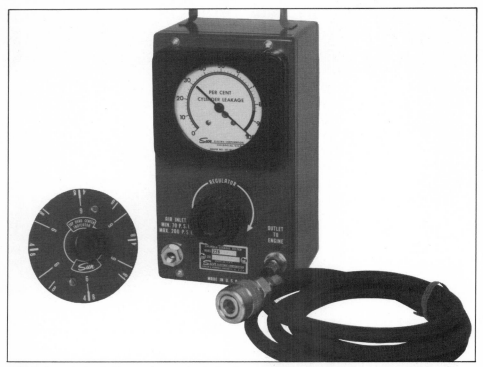

A leak-down tester is the most-accurate device for measuring a cylinder's sealing capability because it eliminates factors which can affect compression-gauge readings. Because leak-down testing is done without cranking the engine, cranking speed, valve timing and cam condition do not affect the readings. Photo courtesy Sun Corporation.

testing is done by most tune-up shops. A leak-down tester is an integral part of Sun Electric's electronic analyzer. Sun recommends that an engine which leaks 20% of its test pressure needs attention, whereas a good cylinder will have 5 to 10% leakage. So, if leak-down testing your engine finds a problem cylinder, you'll know the problem will be rings, valves or a head gasket and possibly a piston or a crack in the cylinder head or the cylinder wall.

Head Gasket or Valves?—Two candidates may be causing the problem at this point—a leaky head gasket or a valve. If two cylinders which are down on pressure and they are adjacent to one another, chances are good the problem is a blown head gasket between the two cylinders. It could also be a head gasket when only one cylinder is affected. If it is a gasket, there will have been an unusual amount of coolant loss from your radiator if cylinder pressure is leaking into your cooling system. This situation is easy to diagnose. Remove your radiator cap and look at the coolant surface when your engine is running and warmed up. If cylinder pressure is escaping into it, you'll see bubbles. Before making this check, make sure your coolant level is up to the mark. If you see bubbles, smell the coolant. If they are caused by a blown gasket, you should be able to detect gasoline or exhaust fumes as the bubbles surface and burst. A sure way to test for this is to take your car to a professional who has a device which "sniffs" the coolant. It indicates whether or not the bubbles are caused by escaping cylinder pressures and are not just recirculating air bubbles. If a bad gasket is found, you'll have to remove a head to replace it. Use the procedures outlined in the engine teardown and assembly chapters for this job. Check the head and block surfaces for flatness. Fix any problems or you may have to repeat the job.

Now for the valves. If you didn't find your compression-loss problem with the piston rings or a head gasket, the last probable customer will be a valve/s. There are numerous reasons for valves leaking, but the cause is always improper seating. A valve may not be fully closing or it may be burned. Both result in an unsealed combustion chamber. If a fully closed valve leaks it's probably burned, so check for full closure first. You'll need vernier calipers or a dial indicatior.

Pull off the valve covers and locate the cylinder you want to check. It should be on TDC (top dead center) of its power stroke—not between its exhaust and intake stroke. This ensures both valves *should* be fully closed. To do this, trace the spark-plug wires of the cylinders you are going to check to the distributor. Put a mark/s on the distributor body in line with the lead on the distributor cap. Remove the cap.

When you crank your engine over and line up the distributor rotor with the mark on the distributor you'll know the cylinder will be reasonably near TDC on the power stroke.

If both valves are fully closed, you should be able to rotate the pushrods with your fingers. It's very unlikely the cam is holding a valve open unless the valves have been recently misadjusted. I know of no instances where the valve adjustment of a small-block Ford has decreased. If adjustment changes, it gets looser. If the pushrod rotates, the rocker arm has unloaded the valve so it is free to close. If it won't rotate, back off that valve's adjustment according to the procedure in the engine assembly chapter.

One word of caution, if your valve train is the same as what was installed at the factory it will have hydraulic lifters, unless you have a HP289 or Boss 302. Hydraulic lifters slightly load the pushrods, making them a little hard, but not impossible to turn. Therefore, don't let this fool you into thinking the valve in question is open. If you have the HP289, Boss 302, or your cam has been changed to a mechanical type, the pushrods will be loose. That's why mechanical cams are noisy. They are loose because the required clearance to ensure closing is about 0.020 inch when hot, as opposed to the zero lash of hydraulic lifters.

CHECKING VALVE LIFT

If all the valves appear to be closing as just described, check to see if the pushrods feel too loose. If one is, it could mean that a valve is sticking in its guide, preventing it from fully closing. To confirm this, you'll have to use your dial indicator to get an accurate reading on actual valve lift. The depth-gauge end of a vernier caliper will work if you don't have a dial indicator, but it's not as easy to use. If you don't have either one of these you can use a 6-inch rule with a slide for measuring depth. It's not accurate, nor is it very easy to use but you've got to use what you have.

Bump the engine over until the valve you're checking is fully opened, compressing the valve spring. Make sure the lifters are fully primed by cranking the engine so you're getting full valve lift. If you have a dial indicator, set the indicator plunger against the top of the spring retainer and in line with the valve stem so you get a true reading, and zero the indicator. If you're using a vernier caliper or a scale, measure from the spring-pad surface—where the spring sits on the cylinder head—to the top of the spring retainer

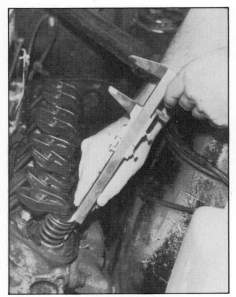
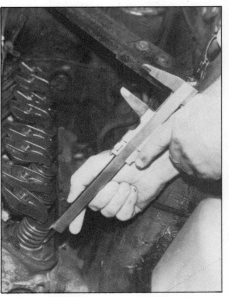
Checking valve lift by measuring how much a valve spring compresses from the valve being fully closed to fully open will confirm whether or not your cam is in good shape or if you have a sticking valve. Valve lift should measure between 0.400 in. and 0.500 in. depending on which engine you have.

Checking lobe lift directly with a pushrod and dial indicator is tedious but accurate and well worth the time if you suspect cam problems.

CAMSHAFT LOBE LIFT (inch)
Originally installed camshaft

Engine	Year	INTAKE At Lifter	INTAKE At Valve	EXHAUST At Lifter	EXHAUST At Valve
221	62–63	0.2375	0.380	0.2375	0.380
260	62–64	0.2375	0.380	0.2375	0.380
289	63–68	0.2303	0.368	0.2375	0.380
HP289	63–67	0.2983	0.457	0.2983	0.457
302	68–75	0.2303	0.368	0.2375	0.380
	*68 only	(0.266)	0.426	(0.266)	0.425
	76–78	0.2373	0.382	0.2474	0.398
Boss 302	69–70	0.290	0.477	0.290	0.477
351W	69–75	0.260	0.418	0.278	0.448
	76–78	0.260	0.418	0.260	0.418

*302s installed in some Fairlanes, Torinos and Mercury car lines.
Lobe lift in parentheses is approximate.

VALVE JOBS AND HIGH-MILEAGE ENGINES

Doing a valve job only on a relatively high-mileage engine may not solve its oil-consumption problem. The reason is, as an engine wears, its parts wear together. As the sealing quality of the valves becomes less, the same thing is happening to the rings and pistons. After the cylinder heads are reconditioned, they will seal better, creating higher compression and vacuum loads on the rings and pistons. Where the rings and pistons were doing a satisfactory job before, they may not be able to seal adequately after a valve job. Increased oil consumption and blowby result. So beware of the valve-job-only solution.

THANKS

The number of years the small-block Ford has been around, plus the changes it has undergone during these years makes it impossible for any one person to write a book such as this. Completeness and accuracy required that I get help from people who make their living selling or manufacturing engine parts, rebuilding engines and servicing them.

At the risk of leaving someone out, I am grateful to: Denny Wyckoff who was always ready to share his years of engine rebuidling experience, and answer one question after another. Daryl Koeppel and his sidekick Jim Hambacher of Holmes Tuttle Ford always took time to answer questions concerning parts, part numbers, interchangeability and those troublesome change levels. They also furnished parts for many of the photos you see throughout the book.

Don Wood and Sam Ellison provided words of wisdom about removing and installing engines. Charlie Camp, drawing on his years of experience as a Ford Service Engineer, was invaluable as he checked the accuracy of my manuscript and made many helpful suggestions. Bill Wheatley and Jeff Quick also took time out of their busy schedules to read over the manuscript and offer suggestions.

All of the photos and drawings credited to Ford Motor Company were cheerfully supplied by Linda Lee of Ford's Parts and Services Division. Others I would like to thank are: Ed Kerchen of Associated Spring, Bob Bub of Cloyes Gear and Products, Bob Lopez of Federal Mogul, Tom Tlusty of Muscle Parts, Cal DeBruin of Sealed Power Corporation, Jack Little of Sun Electric, Terry Davis and Gregg Strugalski of TRW, Dale Cubic of Mr. Gasket, Randy Gillis of Sig Erson Racing Cams, Bill Borrusch and Bob Robertson.

and record the figure. Now, bump the engine over until the pushrod is loose, indicating the valve is closed, and remeasure. You can read valve opening directly with the indicator. Repeat this a couple of times to make sure of your figures if there appears to be a problem with the valve sticking or a worn cam lobe. You should arrive at a valve-lift of 0.368–0.477 inch depending on which engine you have. Refer to the specification on the preceding page for the correct valve lift.

One problem with checking the valve lift on high-mileage hydraulic-lifter engines is the lifters are usually so worn that they can't maintain sufficient pressure to hold the valve completely open. They leak down and partially close the valve even though the lifter is on the toe of the cam lobe. Be aware of this problem if your engine fits this category.

What if a valve lift is not up to specification? You are checking for a sticking valve, so this is the first thing to suspect. As a double check, the installed height of the valve spring when the valve is closed should measure 1.50–1.80 inches. This measurement is made between the spring pad and the *underside* of the spring retainer. Check the specification chart for your engine. If it is less than specified, the valve is sticking or being held open by something. As a final check, back off the rocker-arm adjusting nut until the rocker arm is loose. This will confirm a sticking valve if spring height doesn't change. You can remove the head with some assurance that a valve job is in order. If the spring or valve lift does come up to specification, recheck cylinder pressure, but only after adjusting the valve and warming up the engine. If it comes up to pressure, you've found the source of the trouble, at least with this cylinder.

Checking Camshaft Lobe Wear—If you've found a valve is not lifting to specification, but it's properly adjusted and the installed spring height is right, a worn cam lobe has nothing to do with causing low cylinder pressure, but it has a lot to do with engine power loss.

If your engine is down on power, check all of the valves with the dial indicator. Rather than using the spring retainers to check from, a more accurate method is to check each cam lobe directly via its pushrod. You'll have to loosen all the rocker-arm adjusting nuts and rotate the rockers out of the way of the pushrods. Starting with the number-1 cylinder, mount the dial indicator so it is lined up with the center of the pushrod. You may need a piece of tape around the dial-indicator plunger and pushrod to prevent the pushrod from falling away from the plunger.

Make sure the pushrod is well seated in its lifter and the lifter is solidly against the base circle of the camshaft lobe. With the dial indicator set at zero, slowly bump the engine over with the starter while you watch the indicator. Note the maximum indicated reading and record it with the cylinder number and whether it is an intake or exhaust valve. You'll need this information later for comparison to the other lobes.

Camshaft lobe lift is the difference between the highest portion of the lobe and the diameter of the *base circle* as indicated by the sketch on page 64. Lifts vary from a high of 0.2983 inch to a low of 0.2303 inch, depending on which engine you have. Refer to the specification chart on page 9 for this information.

When checking camshaft lobe wear, you won't have much trouble distinguishing the bad ones from the good ones. When a lobe starts to wear, it goes quickly. It doesn't wear gradually like a cylinder bore. Differences won't be in thousandths, they'll be in tenths of an inch. All the lobes of a camshaft don't all wear down at once, they go one at a time. It's not uncommon to see only one lobe of a cam wear while the rest are perfectly all right.

Regardless of how many lobes you find worn, if you find any, replace the cam and *all* lifters. Otherwise, the new cam will be ruined before you get out of your driveway. Use the procedure outlined in the engine assembly chapter when changing a camshaft and lifters.

One final note before proceeding to the next chapter, if you've discovered a bad cam, but your engine's compression is good and it doesn't use much oil—no more than one quart per 2000 miles, I suggest that you install a new camshaft and lifters. However, if your engine is up to 500 miles per quart and the compression on a few cylinders is less than 75 percent of the highest, it is time to rebuild.

FIRING ORDER

Making assumptions can get you into trouble when it comes to diagnosing, tuning or building an engine. One of the most common assumptions made when it comes to the Ford small blocks is that their firing orders are all the same. Not so, the 351W fires differently: 1-3-7-2-6-5-4-8. The 221 through 302 fires: 1-5-4-2-6-3-7-8. Firing orders are cast into the top of the intake manifolds.

Engine Removal 2

Fully equipped 1975 351W. This largest displacement member of the small-block Ford engine family is destined to become Ford's "big-block." Photo courtesy Ford.

Because you've gotten to this chapter I assume your engine needs rebuilding and must come out. Pulling an engine is one of the most troublesome and potentially dangerous parts of rebuilding an engine if it isn't done right. These troubles are compounded when it comes time to reinstall the engine. A careful and orderly removal job avoids problems, or minimizes them at *both* ends of the project.

PREPARATION
Removing an engine is like diagnosing one—to do it right you'll need special equipment other than the standard set of tools. You should have something to drain engine liquids into, a jack and jack stands to raise and support your vehicle, a method of lifting your engine, a couple of fender protectors and some masking tape to identify loose ends so you can tell where they go when it's time to reinstall your newly rebuilt engine.

Before starting the engine removal process, ask yourself some questions. First, is your lifting device and what it will be attached to strong enough to handle 650 lbs.? A chain hoist hanging from a two-by-four isn't strong enough. Will you be able to leave your car where the engine is removed? Finally, will you be able to move the engine or car once the engine is lifted out of its compartment?

One of the more common methods of pulling an engine is to do it in a garage with a chain hoist attached to a crossbeam. The car is jacked up and supported by jack stands or driven up some ramps, followed by getting the engine ready for removing. After the engine is hoisted out of the engine compartment, the car is set back down on the ground and rolled outside for parking in an out-of-the-way place. The engine can then be lowered to the floor for teardown. The order is reversed at installation time. If ramps are used, it is tough to push a car back up the ramps.

Unfortunately, a too-common result of this method is the money saved by doing the rebuild job by yourself can be negated by the expense involved in rebuilding the garage roof—or worse yet, paying the hospital bills. If the chain hoist support is strong enough, the drawback with this approach is you'll have to move your car before lowering the engine—and once it is down you'll have to move it. A final word of caution about using this method: I've never ever found a passenger-car garage with a beam sturdy enough to support an engine safely.

Then there is the "shade-tree" approach. Set up an A-frame made from 12- to 15-ft.-long, 5- or 6-in.-diameter locust poles—preferably under a tree for shade, of course. Chain them together securely at the top and hang a chain hoist from the chain. Drive your car up two ramps located so the engine ends up directly under the chain fall. Block the car so it won't roll back down the ramp about the time you start to pull the engine. Get the engine ready for removal and lift it up and clear of the car, then roll the car off the ramps and lower the engine to the ground. Even though I'm jesting about the shade-tree approach, an A-frame like I've just described is a lot stronger than a garage beam. With today's lumber prices, you'll probably want to take the next approach. The easiest way I know of to remove engines is with a "cherry picker" rented from your nearest "A-to-Z" rental on a daily basis. They are worth every cent—about $10 a day. Most of them can be towed behind a car. You'll need one for an hour or so when removing, and again when installing your engine. This neat device lets you lift your engine out and move it where you want to start your teardown.

With my sermon over, let's get on with getting your engine out. The small-block series of engines started as a 221CID installed in the 1962 Fairlane and has since that time been installed in all types and descriptions of vehicles. Therefore, due to the complexities involved in explaining how to pull an engine out of every vehicle these engines have been used in,

11

Small-block Fords have been installed in virtually every type of vehicle going. Here is a custom-installed '64 289 in a 1954 Jeep.

I'll generalize and let your common sense fill the voids. The vehicle I used as an example is a 1968 Mustang using a 302 loaded with accessories. About the only thing it didn't have was an air-injection pump.

Before you immobilize your vehicle, clean your engine and transmission to remove as much dirt and grease as possible. The most effective and simplest way of doing this is with a can of spray degreaser and some water. If your car is running, take it to a car wash and just follow the directions on the can. Or, use their degreasing spray if one is available. Otherwise, do it at home with a garden hose. Take fair warning, what's on your engine ends up under the car after you're finished, so act accordingly. After you've finished your engine it won't look like new, but it will be a whole lot easier to work on.

Now, with your engine and transmission clean and everything ready to go, I'll explain how to remove the engine point-by-point:

Fender Protector—Put a fender protector, or a suitable facsimile over each front fender to protect the finish. In addition, they are comfortable to lean on and your tools won't slide off as easily.

Battery—If you have a standard transmission, remove the battery and store it in a safe place. If you have an automatic transmission, disconnect the ground cable at the battery, but leave the battery in because it will be handy later on. Disconnect the coil-to-distributor lead—the

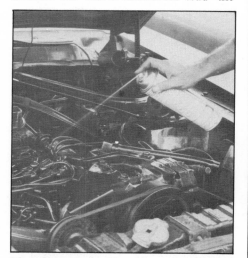
It's easier to work with a clean engine. A can of spray degreaser and your garden hose will do the job. Cover the carburetor to prevent filling the engine and carburetor with water.

one coming out of the center of the distributor cap, or the distributor-to-amplifier lead.

Remove the Hood—If you have an engine compartment light, disconnect it first. Before loosening the bolts, mark the hinge locations relative to the hood. Do this by tracing around the hinges at the hood with a scriber, grease pencil or a piece of chalk. A neat method of guaranteeing that a hood goes back in exactly the same position as it was *before* its hinge bolts are loosened is to drill an 1/8-in. hole up through both hinges and the hood inner panel—*don't go through the hood!* To reinstall the hood, bolt it loosely to the hinges, insert an ice pick through the holes in the hinge and hood to align it, then tighten the bolts. The result—perfect alignment. This will save you the trouble of readjusting the hood when it is replaced.

Remove the front hinge bolts and loosen the back ones while supporting the hood. A helping hand comes in handy—you on one side and him on the other. Remove the back bolts and lift the hood off. Place it out of the way where it won't get damaged. Stand it up in the garage and wire the latch to a nail driven in a wall stud. This will keep it from falling over. Another trick is to put it on the roof of your car. Protect the paint or vinyl top by putting something between the rear corners of the hood and the hood latch and the roof. I assume your car will be parked inside out of the wind, otherwise don't try the hood-on-the-roof trick.

> **PLAN AHEAD**
> It's very tempting and easy to "go crazy" disconnecting everything in sight without regard to where things have to go later on. *Don't rely on your memory.* Even if there are only 5 hoses, there are 120 ways they could be installed and 119 are wrong! Mark *all* the items you disconnect such as wires and vacuum hoses. Wrap masking tape around the end of the wires or hoses and make a little "flag" for writing where the ends go. Put similar flags on the connection itself so you can get it all back together. Put your camera to use. Periodically snap some pictures of your engine from a couple of angles as you disconnect things. A picture *is* worth a thousand words. Finally, use coffee cans or small boxes for the bolts, nuts and washers. If your engine is loaded with accessories, you'll find it especially helpful to label the containers as to the particular unit the parts came from.

Begin the engine removal process by removing the hood and disconnecting the battery ground lead. If you have a standard transmission, remove the battery too.

  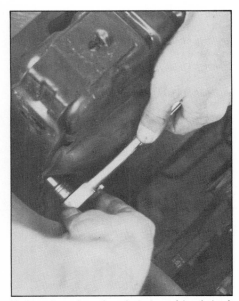

Drain the coolant—loosen the radiator cap so it will drain quicker—remove the upper and lower radiator hoses.

If there is a shroud, unbolt it and lay it back over the front of your engine behind the fan.

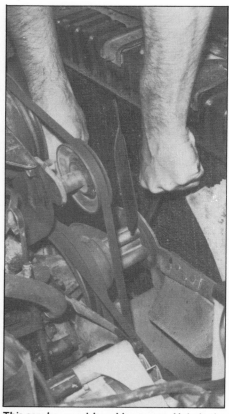

This can be a real knuckle-scraper. Unbolt the fan and spacer from the water-pump flange and lift them out as a loose assembly. Take the shroud out next. This broken shroud must be replaced to ensure good cooling.

Remove the Radiator First—Turn your attention to the fan and radiator. Removing the radiator gives better access to the front of your engine for removing the accessories and prevents the radiator from being damaged during engine removal. One little nudge from the engine as it is being pulled can junk a radiator.

The first impulse is to remove the fan before the radiator, but this is a sure way to remove the skin from your knuckles. The core fins put the radiator in the same family as cheese graters. Drain the radiator—it's faster with the cap off. Remove the top and bottom hoses and replace them if they are over 2-years old, particularly if you live in a hot-dry climate like the southwest U. S. Disconnect the automatic-transmission cooling lines which run to the bottom of down-flow radiators and to the side of cross-flow radiators. Use a tube-nut wrench to prevent rounding off the nuts. This looks like a wide six-point box-end wrench with one flat cut out so you can slip the wrench over the tubing and around the nut. After the nuts are loose, slide them back from the ends of the tubes and connect the two lines with a hose to prevent transmission fluid from siphoning out and messing up your driveway or garage floor. A clean hose the size of your radiator overflow will do nicely, in fact you can use it if it's clean.

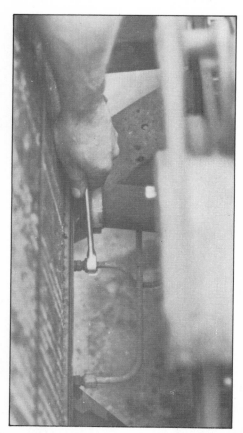
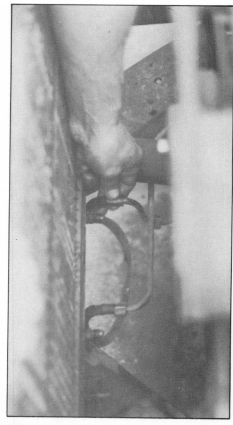

Loosen the transmission-cooling lines from the radiator. Use a tube-nut wrench to avoid rounding off the nuts. Immediately after disconnecting the lines, connect them with a rubber hose about the size of the radiator overflow hose to prevent automatic-transmission fluid loss by siphoning.

If there is a fan shroud, unbolt it from the radiator and lay it back around the fan on the engine. Remove the radiator. First, cut a piece of cardboard to fit the backside of the radiator core, then tape it in place. This will protect the delicate radiator fins *and* your knuckles. If your radiator is bolted solidly to the radiator support, there are 4 bolts, 2 at each side. When you remove the bolts, be ready to support the radiator as you lift it out. Do it with care because fins bend easily if they are bumped. If you have the rubber-mounted type radiator, remove the bracket/s which clamp over the top and lift the radiator out. While you are still holding it, store the radiator in a safe place where it can stay until you're ready to reinstall it. The trunk is usually a good spot. Remove the fan shroud.

Fan—With the radiator out you have a much clearer view of the front of your engine. Loosen the fan-attaching bolts and lift fan, spacer and bolts out together. On clutch-drive fans, use an open-end wrench to get to the bolts between the fan and pulley.

Air Cleaner—Remove the air-cleaner assembly after disconnecting anything which attaches to it such as the hot-air duct, fresh-air duct and the crankcase vent hose.

Throttle Linkage—It will either be the rod-and-lever type or the cable type. A rod type may have an intermediate bellcrank assembly on the intake manifold. Remove it completely from the intake manifold—after disconnecting the rod which runs from the bellcrank to the carburetor. This type has an extra rod from the bellcrank to the dash panel (firewall). Either rotate the rod and bellcrank assembly up out of the way or remove it completely. If the rod runs from the pedal shaft to the carburetor, disconnect it at the carburetor end and swing it up out of the way. On cable types, disconnect the cable at the carburetor end and pinch the tabs or remove the retaining screw at the end of the cable conduit so the cable assembly can be withdrawn from its manifold bracket. Cable linkages used with an automatic transmission require disconnecting the TV rod, or transmission kick-down rod. Disconnect it at the carburetor and wire it to the dash panel.

Don't Trust Your Memory—Here's where a camera and masking tape can be put to use. Label each hose and wire *before* removing it. Disconnect all the hoses and wires from the top of the engine. There'll be a hose from the power-brake booster to the intake manifold and many smaller vacuum hoses, depending on the year and how your car is equipped. An engine-wiring harness usually lays along the inside flange of the left valve cover, retained in clips under three of the valve-cover bolts.

 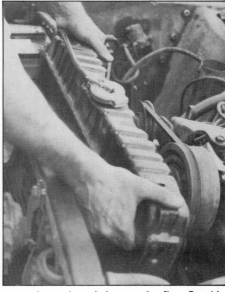

Unbolt the radiator and lift it out, being careful not to bump it and damage the fins. Consider taking it to a radiator shop to have it cleaned.

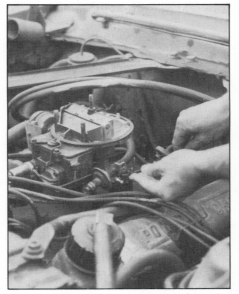 

Disconnect hoses, linkages and wires from the top of the engine, label and tie them out of the way.

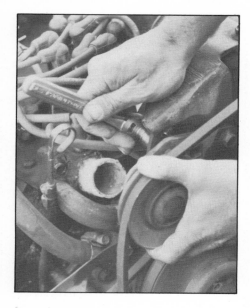

A/C combined with power steering is complicated. Loosen and remove the drive belts. If yours is equipped with a fixed idler pulley like the one pictured below the A/C compressor, remove it first. Then remove the adjustable idler pulley and the compressor-support bracket.

Disconnect the harness leads from the oil-pressure sending unit, water-temperature sending unit, coil, A/C compressor clutch and any emissions-related devices. Bend the clips to release the harness and lay it out of the way after everything is disconnected. Disconnect the heater hoses. Plan to replace them if they are more than 2-years old. You will need to measure them for replacements—5/8- or 3/4-in. diameter and how long?

Pulleys and Belts—Removing the accessories is the most difficult part of removing and replacing an engine. Loosen all the belts and remove them. This will free the water-pump drive pulley, but you won't be able to remove it if your engine has a three-belt accessory drive system. You'll have to remove the crank pulley. After removing the crank-pulley bolts, tap the pulley lightly with a rubber or plastic hammer to loosen it. The pulley pilots into the center of the crankshaft damper, so it may be a little tight. The water-pump pulley will now be free to come off.

A/C Compressor and Power-Steering Pump Go Together—I'll describe an engine which has a complete array of accessories, so disregard those areas which don't apply to your engine. Start with the A/C compressor and power-steering pump. Remove the A/C-idler pulley and bracket assembly attached to the front of the top compressor bracket. With the pulley out of the way, remove this bracket. Some of the bolts are hidden under and behind the bracket. Before removing the A/C compressor, you have to remove the power-steering pump because it uses the bottom compressor bracket for mounting. Remove the pump and its front mount, being careful to support the pump right-side up to prevent fluid loss. Leave the hoses connected to the pump and set the pump aside. Wire it to the left-front fender apron to keep it from falling over and spilling the fluid.

Avoid Disconnecting the A/C Hoses—Here's where you can save some trouble. Instead of disconnecting the A/C compressor hoses so you can remove the compressor from the engine compartment completely, set it aside like the power-steering pump. This eliminates the need to recharge the A/C. Unbolt the bracket from the front of the engine and lay the compressor over to the side. To get the compressor and lines completely out of the way, support the compressor from the left fender. Use a strong cord and bent nail to hook in the wheel opening. With the cord tied to the nail and a bolt threaded into the compressor, hook the nail around the wheel-opening flange with a rag protecting the fender. Tie the cord short enough to support the compressor high enough so it will be out of your way. Remove the bracket from the bottom of the compressor so it won't interfere with removing the engine.

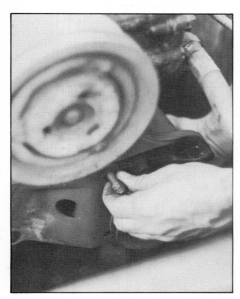
Remove the power-steering pump. You may have to disconnect the pump-to-reservoir return line to get it out of the way against the left fender apron. Make sure the pump sits upright so fluid doesn't run out.

When unbolting the A/C compressor and its mounting bracket from the engine, be ready to handle about 25 pounds.

SAVING YOUR AIR CONDITIONING REFRIGERANT

If you have to remove the A/C compressor prior to removing your engine, the hoses must be disconnected. Fortunately, this can be accomplished without losing all the refrigerant on earlier models equipped with *service valves*. Unfortunately, service valves were phased out during the mid-70s. The valves are in line with the hoses or lines and are usually covered with cadmium-plated caps. If yours is, remove the caps and close the valves. Close (turn counterclockwise) the high- and low-pressure valves at the compressor to isolate the charged system from the compressor. The system is now isolated from the compressor. *Before* removing the hoses from the compressor, loosen the gauge-port cap a *small* amount until you are certain *all* the pressure has escaped. **CAUTION: High pressure in these lines can hurt or blind you if you are careless. Wear goggles to protect your eyes, even when bleeding pressure from the compressor.** Now you can remove the hoses by disconnecting the service valves from the compressor so they stay with the hoses and seal the system. *Don't remove the valves from the hoses.* Place the hoses out of the way and you're ready to remove the compressor after sealing it so it will stay clean and dry. Use the caps which are used to seal new compressors before they are installed. You should be able to pick some up for nothing at your local garage. Remove the compressor with its bracket and set it aside.

A strong cord tied to a bent nail supports the A/C compressor out of the way, subsequently avoiding the need for disconnecting compressor lines and recharging the A/C system. A rag keeps the nail from scratching the fender.

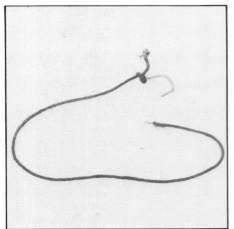

With the left side stripped of its accessories you can remove the fuel pump. Disconnect the fuel line at the carburetor and remove the fuel pump and fuel line together.

A gauge-type oil-pressure sending unit such as this one should be removed now to avoid breaking it off at the block. First remove the oil filter so the extension and sending unit can be unthreaded. Keep your wrench close to the block end of the extension.

and the release-lever pushrod will swing free. You can now remove the bellcrank-and-pushrod assembly. If you have the cable-type linkage, you're in luck because it doesn't have to be removed or disconnected because it isn't attached to the engine.

Remove the Fuel Pump and Line—Disconnect the fuel-tank-to-fuel-pump line at the fuel pump. To prevent siphoning and the consequent fire hazard created by spilled gasoline, push a 3/8-in. diameter bolt into the end of the hose. Make sure the bolt isn't fully threaded. If it is, fuel can leak out around the threads, so it should have at least 1/2-in. unthreaded shank. While you have your head down in there, remove the pump. Disconnect the fuel line at the carburetor by removing the short hose. Use a tubing wrench on the fitting at the pump.

Be Careful Removing the Oil-Pressure Sender—With the fuel pump out of the way, you can see the oil-pressure sending

Alternator or Generator and Air Pump—The left side of your engine should be bare so shift your attention to the other side. There will be an alternator or generator and possibly an air pump. One will be mounted above the other, but which one depends on the year and model. If yours has a generator, it won't have an air pump because they didn't exist back then. If there is an air pump, remove it after disconnecting the hose. Also remove the by-pass valve and other related hardware. As for the alternator/generator, don't remove it as it'll have to be in place for just a little while longer. You can remove its adjusting bracket and loosen the long mounting bolt.

Don't Forget the Filler Tube and Ground Straps—Two easy things to overlook until your engine ceases to move as you're pulling it out are the transmission-filler tube (automatics) and the engine ground straps. These are attached to the right cylinder head at the rear. You might find it easier to remove the ground strap from the firewall. If you do, remember to remove it from the engine and put it back on the firewall after the engine is out. This prevents the strap from getting lost or forgotten between now and when you replace the engine.

Clutch Linkage—With a standard transmission using a rod-and-lever clutch linkage, you have some more up-top work. Some models have a spring from the top of the equalizer bar to the firewall—remove it. Parallel to the spring is a pushrod extending through the firewall to the

Use a penetrant such as CRC® on the exhaust-manifold studs and nuts to loosen them up. Remove the top ones from above. Get the bottom ones from underneath later. It's a tight squeeze both ways.

top of the equalizer bar. Disconnect it at the equalizer, being careful not to lose the bushing. If it looks worn out, count on replacing it. To save it, replace it on the rod and put the clip back in place. The equalizer bar connects the engine to the frame or body. The bar pivots on a bracket attached to the frame or body with 2 bolts. Remove them and the bracket will pull out of the end of the equalizer bar. The equalizer will be free to pull off its pivot at the engine, but you'll have to finish this job from underneath. This can usually be done without raising your car by reaching the bellcrank-to-release lever under the car. Unhook it

unit. It threads directly into the engine block—used with a warning light—or will be on an extension moving the sender outboard away from the front-face of the engine block to accommodate the larger gauge sender. To remove the sender using an extension, you'll have to remove the oil filter because the sending unit threads into the extension at an angle, so the sender swings a wide arc, hitting the filter as the sender-extension assembly is turned. Don't unscrew the sending unit from the extension and then the extension from the engine because this often breaks the extension flush with the engine-block side. Keep your wrench close to the

18

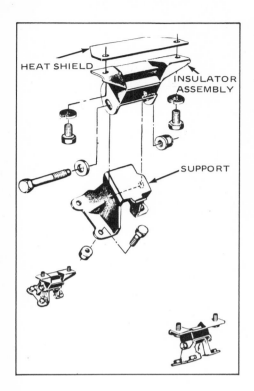

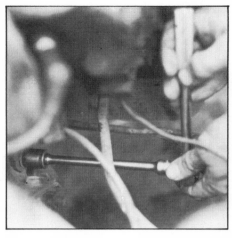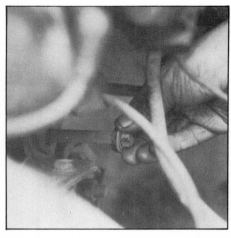

Remove the engine-mount through-bolts by removing their nuts and sliding the bolts and their washers forward out of their mounts. You may have to get at the left side from underneath. Drawing courtesy Ford.

block end of the extension when removing the extension and sender.

The only reason to remove the fuel pump and oil-pressure sending unit from the engine prior to removing an engine is to prevent the possibility of bumping the sender and breaking the extension. If you have the type of sender which doesn't use the extension, you won't have to remove it. This type fits tightly against the engine and the chances of it getting bumped and broken are remote.

Your engine is looking pretty bare on top, the only things left are operations that can be divided between doing from above and below. Before getting the wheels in the air, do these partial jobs to avoid unnecessary climbing up and over the fenders.

Get What You Can From the Top—Remove the top exhaust-manifold exhaust-pipe bolts. These are easier to get to from the top unless you have a *long* extension in your socket set. Some penetrating oil on the threads about an hour prior to removing the nuts and the use of a box-end wrench help avoid damaged knuckles when doing this job—an open-end slips off easily and there's no room for a ratchet handle.

At the back of the engine remove as many bellhousing or converter-housing bolts as you can. You may be able to get a ratchet handle between the firewall and the bolt heads, but not for every bolt.

Engine Mounts—Ford did us a big favor by designing the engine mounts for easy engine removal and installation. Each mount is a two-piece assembly. The rubber portion attaches to the engine and a stamped-steel pad mounts to the frame or body. The two pieces are held together with a *through-bolt* except for the early Fairlane and the high-performance 289s which use vertical bolts or studs. After removing the nut from each through-bolt, the bolts should slide out of their respective mounts very easily because there is no load on the bolts other than the clamping load from torque of the nut on the bolt. To undo the bolt and nut, place a box-end wrench on the nut and use a socket and ratchet with an extension on the other. As you start to unthread the bolt from the nut, the box-end wrench will rotate against the engine, body or frame so you won't have to hold that end. When the nut is off, slide the bolt forward out of the mount. I find it easier to get to the left mount from below. You'll have to assess your particular situation.

Jack Your Car Up—It's time to get your car in the air. A truck usually has enough ground clearance to work underneath without raising it. Not so with a car. A hydraulic floor jack is great to have at this point. You'll only have to raise the front of your car for removing the engine. To do this, place the jack under the number-two crossmember. Frame cars have a substantial crossmember to which the lower control arms of the front suspension attach. Unit-body cars have a less-substantial-looking tubular crossmember which bolts to the body. It is all right to use this crossmember to raise a car, but be careful because it's easy to get the jack pad under the steering linkage, possibly bending the center link. A 2" x 4" wood block between the jack and crossmember helps.

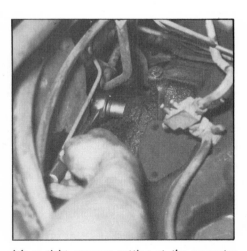

It's a tight squeeze getting at the converter housing or bellhousing bolts from above, but it's easier than from underneath. Get as many as you can from topside before raising your car.

With the car in the air, place the jack stands under the frame or body of the car rather than the front suspension. I prefer this method because it's possible the car may shift sideways. This doesn't usually result in the car falling, but it's not something to chance. As you can see from the photo, I placed the jack stand under the swaybar bracket, a good substantial location which gets the jack stands forward out of your way. If you don't use jack stands to support your car, whatever you use, make sure it is *wide-based*, particularly side-to-side so the car can't shift sideways. Block the back tires so the car can't move backward or forward.

When jacking up the front of a car, it's best to use the crossmember just behind the engine. In the case of a unit-body car such as this Mustang, the bolt-on tubular crossmember will support the load, but don't jack against the steering linkage. You'll bend it.

Support your car solidly on jack stands. Whatever you do, don't depend on a jack or concrete blocks. It could be fatal.

Exhaust-pipe to manifold bolts are hard to reach. A universal socket and very long extension is one answer.

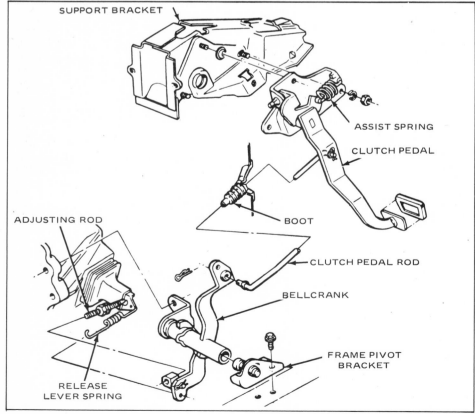
To remove the clutch-linkage bellcrank, remove the pivot bracket from the frame or body siderail, then disconnect the clutch-pedal rod and the release-lever pushrod and spring from the bellcrank. This permits the bellcrank to slide off the pivot at the ending. Drawing courtesy Ford.

Do not use bricks, cement blocks or cinder blocks. These materials work fine with evenly distributed loads, but they crack or crumble when subjected to point loading as in supporting a car. If I seem to be dwelling on this subject too much, when a car falls the result is often fatal—if it doesn't kill someone, the least is usually serious bodily harm. *Be careful.*

Now's the time to get your trusty creeper into service—or a big sheet of cardboard works well, particularly if you're over dirt. A creeper doesn't roll well in the dirt.

Finish the Clutch Linkage—With the car in the air, finish those partially finished jobs you started from the top. If you have a standard transmission with a rod-and-lever clutch linkage, finish removing the equalizer bar and release-lever pushrod assembly. Just remove the retaining spring from the release lever and you can pull the bellcrank off the engine pivot to remove it.

Exhaust System—Finish disconnecting the exhaust system from the manifolds. Here's where a 12–15-in. extension for your socket set comes in handy for reaching the nuts.

Cooling Lines and Starter-Motor Cable—With an automatic transmission, make sure the cooling lines are unclipped from the engine. Some have clips and some don't. If yours does, just push the lines out of the clips to release them. While

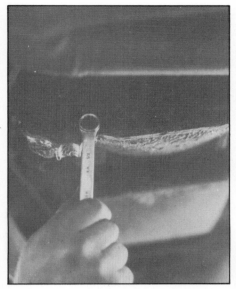  

To disconnect the converter from the flex plate you'll have to reconnect the battery ground. Remove the cover plate to expose the back-side of the flexplate, then have a friend "bump" the engine over to expose each attaching nut with the understanding that he is not to touch the starter switch until you've given the word. Or, use a remote starter switch to bump the engine over. A wrench on the crankshaft damper bolt also works.

you're looking up where these lines go past the engine—usually on the right side near the top of the oil pan—make sure the starter-to-battery lead is out of its bracket. The bracket is on the right-front engine-mount-to-engine bolt. Bend the bracket loop and slide the cable and its grommet out.

Engine Mounts—If you couldn't reach the left engine mount from topside to remove the bolt, do it now. Remember, once the nut is off, just slide the bolt out.

Expose the Converter Attachments—With an automatic transmission, remove the converter cover at the front of the converter housing. Once you have the 2 attaching bolts out, remove the bolt through the engine plate into the housing on the opposite side from the starter. After this is out, slide the cover from between the engine plate and converter housing.

Get Power to the Starter—Reconnect the battery ground so you'll have power to the starter. Now, if you don't possess a remote starter have a trustworthy friend *bump* the engine over to expose each of the converter-to-flexplate nuts for removal—a socket on the front of the crank pulley will accomplish the job just as effectively. CAUTION—If you use a friend; make sure he understands the ignition switch *is not* to be touched without your direction.

Now for the Starter Motor—After removing the converter nuts, disconnect the battery cables at the battery. Disconnect the starter cable and remove the starter. Remove the bottom starter bolt first and then the top one while supporting the front of the starter with one hand. Lifting the starter out is tricky on unit-body cars not equipped with rack-and-pinion steering. Slide the starter forward and drop its nose (geared end) between the steering linkage and the converter housing.

Finish Disconnecting the Bellhousing—Regardless of which transmission you have, the rest of the removal process is pretty much the same. The only job left before lowering your car is to remove the remaining bellhousing/converter housing bolts. After doing this take one last look underneath just to make certain *everything* is disconnected.

Remove the Alternator or Generator—After setting your car down, the last thing to do before pulling the engine is to disconnect the battery ground and the alternator lead from the engine. Remove the alternator/generator and tie it to the side out of the way. Remove the battery.

Lift the Engine Out—Take another last look around the engine compartment to double-check that everything is disconnected—inevitably there will be something. Once that you're confident the engine is ready, position your lifting device over the engine, or the car with the engine under it, whichever. Attach a lifting cable

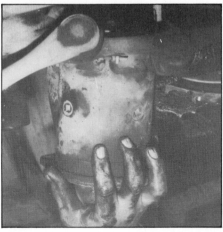

Remove the starter bolts and slide the starter forward out of the bellhousing or converter housing. Some earlier unit-body cars are tight in this area and you'll have to tilt the starter down behind the steering linkage.

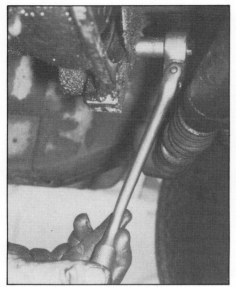
When removing the remaining converter-housing or bellhousing bolts, don't overlook the bolt attaching the engine plate to the housing on the side opposite the starter.

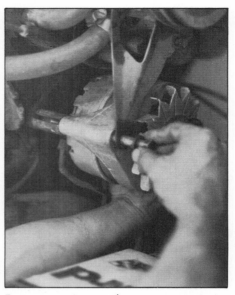
Remove your alternator/generator, remembering to disconnect its engine ground. Position it out of the way. Remove other electrical leads to the engine such as the starter cable and engine ground at the rear of the right cylinder head where the automatic-transmission filler tube is attached.

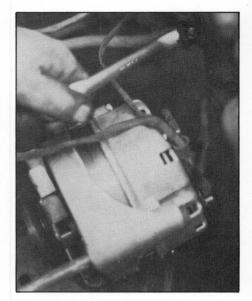

with looped ends or a chain to the front of one head and rear of the other with bolts and a large washer on each bolt. The chain or cable will be positioned diagonally across the engine. Make sure there is enough slack in the cable/chain so you'll be able to set the engine all the way down where you want once it's out of the engine compartment. The chain or cable must be short enough so the engine can be lifted high enough to clear the body work when pulling it out. Hook the chain approximately in the middle so the engine will be balanced once it is free from the transmission and the engine mounts. Put a jack under the transmission so it won't fall once the engine has been pulled away. Now, raise the engine until it can be pulled forward and clear of the engine mounts. Bring the jack in contact with the transmission again and now you're ready to disengage the engine from the transmission. This may require a little juggling. Standard transmissions mean moving the engine farther forward so the transmission-input shaft fully disengages from the clutch. Don't try to force the engine if it hangs up. You could damage something. Check around to see what bolt, ground strap or other connection you may have overlooked. Disconnect it and proceed

With an engine hoist positioned over the engine, attach a chain diagonally across the engine for balance—to the front of one head and to the rear of the other one.

A jack positioned under the transmission prevents it from falling when the engine disengages from it.

With the engine lifted high enough to clear the engine compartment, it can be rolled to a convenient place to begin stripping it down, but first . . .

 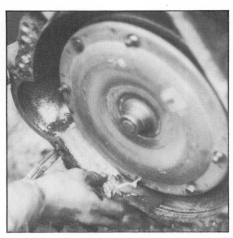

. . .Don is marking the flex plate and converter stud so the two can be reassembled in their original positions.

to remove the engine. As soon as the engine is completely free of the transmission, hoist it out.

Mark the Converter-to-Flex-Plate Location—With light-colored paint—spray or brush—mark the one stud on the torque converter and the matching bottom hole on the engine flex plate. Flex-plate holes and torque-converter studs fit very tightly, so if you mark them to go back the way they came apart, this is one variable you'll know is right if things don't click together when you are reinstalling the engine.

Secure All the Loose Parts—If you don't want to leave your jack under the transmission or want to free your car so it can be rolled, support the transmission so it doesn't fall when you lower the jack. Old coat hangers work well for this job. The same thing goes for the exhaust system.

With your engine out and hanging in mid air, begin stripping it down to the bare essentials as described in the teardown chapter.

Before the parts are scattered, collect and store them in a safe place where they won't get mixed in with your lawn mower, motorcycle, snowmobile or any other parts you have lying around. You'll be glad you did. If you get most of the grease and oil off the parts and put them in greaseproof containers, the inside of your car and in the trunk are good places to keep miscellaneous parts.

Continue with tearing your engine down by skipping the next chapter and go directly to Chapter 4. Save the Parts Identification and Interchange chapter for when you're ready to sit down and relax.

If you have to return the jack or you want your car free to move you'll have to support the transmission and exhaust system. Wire coat hangers are used here.

3 Parts Identification and Interchange

HP289 is truly a high-performance engine producing 271 HP at 6000 RPM. It is characterized by its solid-lifter valve train, free-flowing exhaust manifolds, dual-point centrifugal-advance-only distributor and high-strength nodular-iron crankshaft and main-bearing caps. Photo courtesy Ford.

It's a real money-saver to know what parts can be interchanged, especially when you need to replace one of your engine's major components with parts from another engine.

Will a 289 crankshaft fit and work in a 302, or will 351W heads fit a 302? How can you identify them if they will? Armed with this information you can visit the local junkyard knowing your options rather than being forced to purchase for your specific engine—or having to play Russian Roulette with parts.

You need to know exactly which engine you have before you order basic parts or try to decide what parts will interchange with it. You need to know the *displacement, year* and *change level*. Change level is Ford's way of identifying a change made in the midst of a model year.

IDENTIFICATION

What Engine Do You Have?—To start from ground zero, I'll assume you don't know what engine your car is equipped with. Look at the car's identification tag on the driver's front-door hinge pillar or on the door near the latch for 1967 and prior models and under the left-front corner of the windshield for later models. The five-digit alphanumeric group contains model year and engine-code information. The first number is the year—7 for 67, 77 or 87. You have to know the decade. The last letter is for the engine. You must use both the year *and* the engine code to determine what engine was installed in the car. For example, F indicates a 260-2V or 302-2V. Because they didn't overlap in years produced, the letter F was reused. The letter codes are indicated in the table.

CODE	ENGINE	COMMENTS
L	221-2V	
F	260-2V	Same as 302-2V code
C	289-2V	
A	289-4V	Premium fuel engine
K	HP289	Mechanical camshaft
G	Boss 302	Unique heads and intake manifold
F	302-2V	Standard passenger car engine
D	302-2V	Taxi or police
J	302-4V	Premium fuel engine
G	302-2V	'72 and up Bronco, '69 and up pickup truck Boss 302 also uses G
H	351W	Same code as 351C

If your engine is original equipment you can identify it by the code on the vehicle-identification tag on the left-front door-latch face or hinge pillar for 1967 and earlier models, or on the top of the instrument panel at the far left side. Cross-reference this code to the accompanying list. Drawings courtesy Ford.

Suffixes W and C indicate the city in which the engines are manufactured—Windsor and Cleveland 2V means a two-barrel carburetor. 4V means a 4-barrel carburetor. If your engine is a Cleveland you need another book, because I am not talking about how to rebuild Cleveland engines.

351W and 351C engines have the same codes, so you'll have to know how to tell them apart. The major difference is cylinder-head design. 351W heads are smaller than the 351C. Because you won't

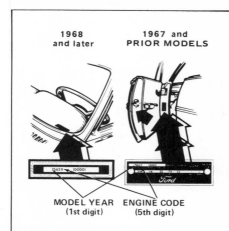

24

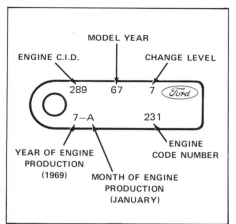
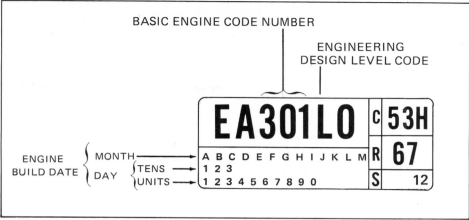

You'll need the information on this tag or decal when ordering parts for your engine.

have one to compare against the other, their relative sizes can't be used for identification. Consequently you need to know their distinguishing features if the engine-identification sticker is not on the air cleaner or there is no air cleaner. The cylinder-head and exhaust-manifold mounting surfaces of the two engines are angled differently. The 351W exhaust-manifold surface is close to vertical; the 351C surface is angled 45° so it's generally in line with the side of the cylinder block. This makes the 351C head wider between the intake-manifold and exhaust-manifold mounting surfaces. Those two surfaces are also parallel. Although the 302 and Boss 302 cylinder-block exterior dimensions are the same, their cylinder heads are as different as the 221–351W heads are different from the 351C. Boss 302 and 351C heads are quite similar, although not the same.

Engine Tag or Decal—Once you've determined which engine you have, you'll also need to know its date of production, change level and engine-code *number*—this is different from the engine's code letter on the vehicle identification tag. The engine-code number used by the Ford parts man when he's ordering parts is on a tag or decal attached to the engine. From January 1964 through February 1973, tags were mounted at the front of the engine, usually sandwiched under the coil-mounting bracket, dipstick-tube bracket or under the water-temperature sending unit. Unfortunately, when a tag is removed for one reason or another, they are usually discarded rather than being replaced—a *big* mistake. Beginning in February 1973, tags were replaced by decals on the front of the right valve cover. The sketch shows how to read a tag or decal.

If an engine's tag or decal is missing, a couple of alphanumeric groups may be stamped on the front of the engine block. The first digit gives you the model year. For example 6J100358 indicates 1966. The other group is the *engine build-date*

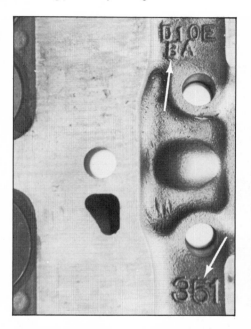

You'll find cylinder-head casting numbers under the intake ports and in the rocker-arm area under the valve cover. In many cases you'll also find engine displacement (arrows). Additionally, the 302 head indicates its carburetor application because of its unique casting. 4-V heads have smaller combustion chambers for higher compression than the 2-V heads.

code, or the date the engine was manufactured. It may look like this: 9A12S. The first digit 9 is calendar year 1969 and the second A is the month January. The letter indicates the month starting with A for January and continuing through M. I is not used.

Casting Numbers—When a component such as a cylinder head or block is cast, a number cast in it is appropriately called a *casting number*. Casting numbers are extremely helpful when identifying an engine or its parts. Unfortunately, casting numbers are not 100-percent accurate because

UNDERSTANDING THE PART NUMBERS

If you are like most people the first time you cast your eyes on a Ford part number, you probably thought it looked unnecessarily complicated. However, this first impression is gradually changed as you understand what all the numbers and letters stand for.

Engineering and Service Numbers—At the risk of confusing you, you should first know that there are two numbers for each *finished part: the engineering,* or *production* number and the *service* number. They are just what they say. The engineering number is assigned by engineering when a part is approved for production. This number is used by the assembly plants. The service number is assigned when the part goes into the part-distribution system. It is a different number because how a part is finished and packaged for service is different from its original production counterpart. It's the number used by your Ford parts man to look for or order. He doesn't want to know about the engineering part number. However, you may need to because the engineering number is the one appearing on many parts.

Casting Numbers—Casting numbers are special engineering numbers placed on a casting to assist in identification by the plant. They are cast on a part at the time of manufacturing, and apply only to the basic casting. One casting can be machined to make slightly different parts, thus creating several different parts, and consequently several different finished engineering- and service-part numbers. Using casting numbers to identify a part is sort of like playing horseshoes. It doesn't count much when you're close, but it does count—and it very well may be the only number you have to work with.

Dissecting the Numbers—Ford part numbers, regardless of whether they are engineering or service numbers, always consist of three distinct groups: *prefix, basic part number* and *suffix.*

Engineering		
Casting number	Production number	Service number
C7OE-6090-C	C7OE-6049-C	C7OZ-6049-G

Prefix—The four-digit *alphanumeric* prefix tells the year the part was released for production, the car line the part was originally released for and by what Ford *engineering division* (chassis, engine, body, etc.) or in the case of a service part, the Ford *car division* the part is for—Ford or Lincoln-Mercury. For example, the above part numbers are for a 289 cylinder head originally released in 1967 as indicated by the first letter—C for the 60s, D for the 70s and so on. The following number is the year in that decade—7. The third digit, usually a letter, indicates the car line the part was originally designed for. Here are most of them.

The O designation is given to 289 engine parts because it is a direct descendant of the 221 specifically designed for use in the Fairlane car. The same applies to many small-block parts. The last digit, or letter in the prefix in this part number indicates the part was released for production by the Engine Division: A is for chassis, B is for body and E is for engine. Makes sense because it's an engine part. This applies to engineering parts, however the digit in a service part refers to the car division—Z for Ford division, Y for Lincoln-Mercury Division and other letters for special parts such as X for the extinct Muscle Parts Program.

The Basic Part Number—Regardless of whether it's an engineering or a service part number the basic part number will be the same. For example, 6049 is for *all* cylinder heads, 6303 is for crankshafts and 6010 is for blocks. The number for the casting used for machining these parts is different. Referring to the cylinder head again, the basic *finished* part number is 6049 whereas its casting number is 6090. Because it is relatively easy to put a casting number on a part while it is being cast, it's the casting number that appears on the part—great for them and terrible for the guy trying to identify a part. Also, the number that appears on a casting may not include the basic casting number for the simple reason that you don't need a number to tell you you are looking at a block or an intake manifold. The number generally consists only of the prefix and the suffix, or C70E-C for the cylinder head.

Suffix—A part-number suffix generally tells you the *change level* of a part, regardless of whether it is applied to the casting, the finished part or the service part. A applies to a part produced as it was originally designed, B indicates it was changed once, C twice and right through the alphabet in sequence, excluding the letter I. When the alphabet has been gone through once, the suffix grows to two letters and starts over as AA, AB, AC and so on. How does a change affect the other two numbers? A service part and its number can change independently of the casting part and its number and the engineering part and its number simply because it comes after these two in the scheme of things. Using the same reasoning, a finished, or engineering part can change independently of the casting, but not of the service part. A casting affects both the finished and the service parts. This is why the suffixes of all three numbers rarely ever match.

VEHICLE IDENTIFICATION CODES

- A— Full-size Ford
- B— Bronco 70–73
  - Maverick 75–77
- C— Torino Elite
- D— Falcon 60–69
  - Maverick 70–74
  - Granada 75 and up
- E— Pinto
- G— Comet 61–68
  - Montego 69–76
- I— Monarch
  - Versailles
- J— Industrial engines
- K— Edsel
  - Comet 60–61 and 75 and up
- L— Lincoln 58–60
  - Mark III, IV, V
- M— Mercury
- O— Fairlane 62–68
  - Torino 69–76
  - LTD II 77 and up
- Q— Fairmont
- S— Thunderbird
- T— Truck
- U— Zephyr
- V— Lincoln 61 and up
- Z— Mustang

castings are frequently machined differently, thus generating different parts. This also generates different *part numbers* which don't appear on the part! To make matters worse, it is conceivable these parts won't interchange. Consequently, a part's casting number in conjunction with its physical design must be used to identify it. So let's get on with the physical make-up of the various engines and their parts.

BLOCKS

221 and 260 CID—These blocks are virtually the same except being cast differently to provide different bore sizes—3.50 in. versus 3.80 in. Due to their "thin-wall" design, the 221 cannot be bored to accept 260 pistons, nor the 260 to accept 289 pistons. The major differences in the 221-260 CID engine series are between those produced prior to and after February '63. Early engine-mount bosses and their holes are 6 inches between centers; later ones were increased to 7 inches. If you don't feel like measuring, early 221 and 260 blocks had 2 *freeze*, or *core* plugs per side. A third one was added when the engine mounts were widened. If you can't find the block with the same engine-mount spacing you can make the other one work simply by substituting engine mounts. To identify the engines by casting number use the accompanying chart. You'll find numbers on 221 engines at the rear of the block under the intake manifold. All others are located under the right cylinder bank at the rear.

289 CID—A new casting was required to increase the Ford small-block to an even 4.00-in. bore to obtain 289 cubic inches while maintaining the 2.87-in. stroke of the 221 and 260 engines. Like the 221 and 260 engines, early 289s up to mid-August '64—used a five-bolt bellhousing pattern. It was then changed to accept a larger six-bolt bellhousing. Watch this when looking for a 289 block. The bellhousings are *not interchangeable*. You'll find the casting numbers at the rear of the right cylinder bank. Beginning with the 289, the engine displacement is cast in the block between the lifter bores for easier identification.

HP289—This block is similar to the standard 289 block with one major exception. The main-bearing caps are much beefier and are cast from high-nodular iron. If you want a small-block with a tough *bottom end* this is the one to look for. The problem is they are rare and expensive. This brings up an important point. When purchasing an engine block *make sure you get its main-bearing caps and bolts too*. Particularly the caps. This avoids the necessary and expensive *line boring* job required when installing different bearing caps on a block. The HP 289 was color coded with an orange paint swatch on the rear face of block, or in front of the flywheel or flexplate to distinguish it from the standard 289 block.

302 CID—The 302 block is basically the same as the 289 block. They are fully interchangeable except they will not fit the five-bolt bellhousing of the early 289. You do have to bolt your engine to a transmission. Identifying a 302 is the same as a 289. Its casting number is at the back side of the right cylinder bank and 302 is cast between the lifter bores.

Boss 302—There are two major differences between this engine and standard 289/302 blocks. The Boss 302 has four-bolt main-bearing caps, making its bottom end very strong. Its cylinder walls are slightly thicker.

The first 302 Boss engines produced in 1969—with C9ZE casting number—had a severe cylinder-wall-cracking problem.

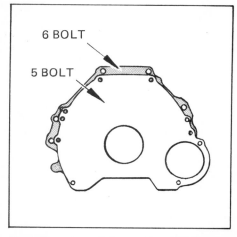

Overlaid early and late engine-plate outlines illustrate difference between the pre '65-1/2 5-bolt and later 6-bolt bellhousing patterns.

Boss 302 4-bolt main-bearing caps make the bottom end of this engine the strongest one in the small-block family. The screw-in core plugs (arrow) are further indication of its racing heritage.

The fastest way to correct the problem was to put the racing Boss 302 into production—casting C8FE—with slightly thicker and ribbed cylinder walls. It was an expensive proposition for Ford, but not what it would've been had they waited for a production fix to be incorporated and then have to replace many of the engines produced up to the fix. The 1970 model-year casting—D0ZE—incorporated these fixes. A Boss 302 engine was released for 1971 production, however Ford's sudden and unexpected disassociation with anything having to do with high performance stopped the Boss 302 in its tracks at the end of the 1970 model year. The only thing left of this engine is block casting D1ZE, the service block for all Boss 302s. These blocks can be identified by their casting numbers on the right cylinder-wall bank.

One minor exterior difference between the Boss 302 block and the others is they use six screw-in plugs to close the core holes in the sides of the block rather than the sheet-metal, cup-type plugs.

CYLINDER-HEAD CASTING NUMBERS & SPECIFICATIONS (BASIC CASTING 6090)					
ENGINE	YEAR	CASTING NUMBER	COMBUSTION-CHAMBER VOLUME (CC)	VALVE DIAMETER (INCHES) INTAKE/EXHAUST	COMMENTS
221	62-63	C2OE-A, B, C, D, E C3OE-A	45	1.59/1.39	
260	62-63	C2OE-F C3OE-B	54.5	1.59/1.39	
	64	C4OE-B	54.5	1.67/1.45	
289	63-64	C3AE-F C3OE-E, F C4AE-C	54.5	1.67/1.45	
	65-67	C5DE-B C6DE-G * C6OE-C, E C6OE-M * C7OE-A * C7OE-B C7OE-C * C7OZ-B * C7ZE-A	54.5	1.67/1.45	* May be machined to accept air pump plumbing. Rail-type rocker arms.
	68	* C8OE-D, L, M	63	1.67/1.45	
HP289	63	C3OE	49.2	1.67/1.45	Machined to accept screw-in studs.
	64-67	C4OE-B C5OE-A C5AE-E	54.5	1.78/1.45	
302	68-70	C8OE-F C7OE-C, G C8AE-J C8DE-F C8OE-J, M * C8OE-K, L C9TE-C D0OE-B	53.5 63 58.2	1.78/1.45	Used with 4-V engine.
	71-74	D1TZ-A D2OE-BA			
	75-76	D5OE-GA, -A3A, -A3B	58.2	1.78/1.45	Internal air-injection ports.
	77	D7OE-DA	69	1.78/1.45	
	78	D8OE-AB	69	1.78/1.45	Common with 351W, uses stamped-steel rocker arms.
Boss 302	69	C9ZE-A	63	2.23/1.71	
	70	D0ZE-A D1ZE-A	58	2.19/1.71	
351W	69-74	C9OE-B, D D0OE-C, G D0OZ-C	60.4	1.84/1.54	
	75-77	D5TE-EB D5AE-CA	60.4	1.84/1.54	Internal air-injection ports.
	78	D8OE-AB	69	1.78/1.45	Common with 302, uses stamped-steel rocker arms.

CYLINDER-BLOCK CASTING NUMBERS & SPECIFICATIONS (BASIC CASTING 6015)				
ENGINE	YEAR	CASTING NUMBER	BORE (INCHES)	DECK HEIGHT (INCHES)	COMMENTS
221	62	C2OE-G C3OE	3.50	8.206	5-bolt bellhousing pattern, 7/8 inch between bores; 6 inches between motor-mount holes, 2 freeze plugs per side, casting no. on top-rear of block.
	63	C3OE-A			7 inches between motor-mount holes, 3 freeze plugs per side, casting no. under right-rear cylinder bank.
260	62	C3OE-B	3.80	8.206	5-bolt bellhousing pattern, 9/16 inch between bores, 2 freeze plugs per side, 6 inches between motor-mount holes, casting no. on top-rear of block.
	63-65	C3OE-C C4OE-B, -D, -E	3.80	8.206	7 ins. between motor-mount holes, 3 freeze plugs per side, casting no. under right-rear cylinder bank.
289	63-64	C3AE-N C4OE-C, -F C4AE C4DE	4.00	8.206	5-bolt bellhousing pattern.
	65-68	C5AE-E C5OE-A C6AE-C	4.00	8.206	6-bolt bellhousing pattern.
HP289	63-64	C3OE-B C4OE-B	4.00	8.206	5-bolt bellhousing pattern, uses thicker than standard main-bearing caps, orange color code on rear of block.
	65-68	C5AE-E	4.00	8.206	6-bolt bellhousing pattern, orange color code on rear of block.
302	68-69	C8OE-A, -B C8TE-B	4.00	8.206	6-bolt bellhousing pattern.
	70-74	C8TE-B D1TZ-E D1OE-AA D4DE-AA	4.00	* 8.206	* 73-76 deck height is 8.229". 6-bolt bellhousing pattern.
	74-77	D4DE-AA 75ZY-AA	4.00	* 8.206	6-bolt bellhousing pattern.
Boss 302	69	C9ZE C8FE	4.00	8.209	6-bolt bellhousing pattern.
	70	D0ZE-B	4.00	8.209	6-bolt bellhousing pattern.
351W	69-70	C9OE-B	4.00	9.480	6-bolt bellhousing pattern.
	71-74	D2AE-BA D4AE-DA	4.00	* 9.503	* 71-73 deck height is 9.480".
	75-77	D4AE-AA	4.00	9.503	6-bolt bellhousing pattern.

CRANKSHAFT CASTING NUMBERS & SPECIFICATIONS (BASIC CASTING 6303)						
ENGINE	YEAR	CASTING NUMBER	STROKE (INCHES)	JOURNAL DIAMETER (INCHES)		COMMENTS
				MAIN	CONNECTING ROD	
221, 260, 289	62-68	C2OZ C3OZ IM IMA	2.87	2.2486	2.1232	
HP 289	63-67	IM	2.87	2.4286	2.1232	High-nodular cast iron, orange color code, last counterweight may be polished for Brinell hardness testing, letter K stamped on next to last counterweight.
302	68-69	C8AZ-A 2M	3.00	2.2486	2.1232	
	70-74	C8AZ-A 2M 2MA	3.00	2.2486	2.1232	
	75-78	2M 2MA	3.00	2.2486	2.1232	
Boss 302	69-70	D0ZE-A 7FE-B	3.00	2.2486	2.1232	Forged steel, '69 crank is cross-drilled and throws hollowed.
351W	69-72	3M 3C	3.50	3.0000	2.3110	
	73-78	3MA	3.50	3.0000	2.3110	

Casting numbers listed above and on preceding page are correct according to the best available information. Use physical dimensions to make final parts identification.

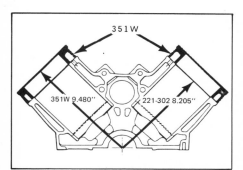

Drawing illustrates 0.275-in. higher deck height of the 351W compared to the 221—302 blocks. Drawing courtesy Ford.

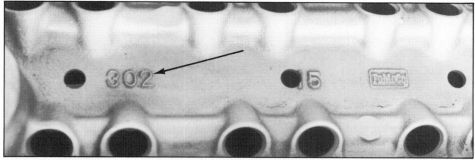

Engine displacement is cast between the lifter bores (arrow) and the block casting number is under a ledge on the side of the right-rear cylinder bank (arrow).

351W—The 351W engine marks the first major change of the small-block Ford's block. The *deck height*—distance from the center of the crankshaft main-bearing journals to the cylinder-head gasket surface—was raised 1.274 in. to accommodate an additional 1/2-in. stroke. Deck height was increased from 8.206 in. to 9.480 in. The 4.00-in. bore was retained. To handle the additional bearing load resulting from the increased engine dis-

Eccentric weight on this crankshaft damper is part of the external weight used to balance the small-block Ford. The rest of the external balance weight is on the flywheel or flex plate.

placement, main-bearing diameters were increased from 2.25 to 3.00 in. These changes prevent interchanging the 351W with any of the other small blocks. Casting numbers are located the same as the 289 and 302 engines.

CRANKSHAFTS

One method used to achieve such a light compact engine was to limit the size of the crankshaft counterweights *inside* the crankcase. Consequently, additional counterweighting had to be provided *outside* the crankcase to balance the engine. Counterweights were added to the crankshaft damper, or harmonic balancer and to the flywheel or flexplate. This makes all small-block Fords *externally balanced*. Other than the Boss 302, another common characteristic of 221, 260, 289, HP289 and 351W crankshafts is the material used to produce them—nodular iron. One slight difference here is the HP289 crank was produced using iron with higher *nodularity*. This simply translates into a stronger crankshaft, in keeping with the HP289s stronger main-bearing caps. The Boss crank goes one step further. It is machined from forged steel, and the original 1969 crankshaft is *cross-drilled* for improved lubrication. Its con-

necting-rod throws are hollowed out for lighter weight. Both operations were dropped for the 1970 model year as a cost-saving measure, consequently the '69 crank is a better part.

With exception of the 351W, all main- and connecting-rod-journal diameters are the same at 2.2486 in. and 2.1232 in. respectively. 351W crankshaft journal diameters were increased to 3.000 in. and 2.311 in. respectively. Strokes remained the same 2.87 in. from the 221 through the 289. With the 302s stroke increased to an even 3.00 in., its crank *throw* was increased 0.065 in. A crankshaft's *throw* is the center distance between the main- and connecting-rod journals. It is *half* the stroke. The 351W

You'll find the crankshaft casting number on the side of the front throw. 2M indicates this is a 302 crankshaft.

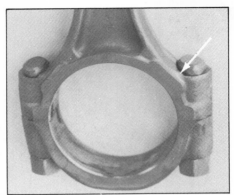
3/8-in. rod bolts and spot-faced seats make Boss 302 connecting rods much stronger than standard 5/16-in. bolted broached-seat rod at right. The HP289 rods use the larger bolts, but rods are broached.

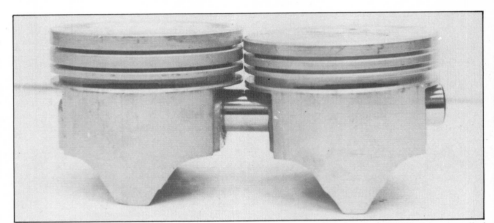
Compression height of a 351W piston is 0.136 in. higher than a 302 piston. A 351W piston installed in a 302 block will protrude above the gasket surface.

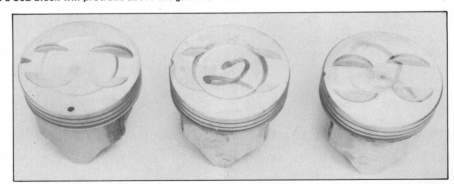
These 302 pistons illustrate changes in compression ratios and clearance volumes over the years. Left to right, the '68—'72 piston is slightly dished whereas the '73—'76 piston is more so, resulting in a one-point lower compression ratio with the same cylinder head. The 302/351W cylinder head with its large combustion chamber requires a flat-top piston to achieve an 8:1 compression ratio.

achieved its additional displacement by a 1/4-inch throw increase, giving a 3.50-in. stroke. A concurrent 1.274-in. increase in the 351Ws block deck height accommodated longer rods and pistons with more compression height.

Crankshaft Replacement—If your connecting rods and pistons appear to be in good shape, but your crank can't be reused, you'll have to replace it with a crank having the *same stroke and journal diameters* as the one originally installed in your engine. If your engine displaces 221, 260 or 289 cubic inches you can use a crank from any of these engines because they all have the same stroke and journal sizes. Not so with the 302, Boss 302 or 351W engines. A 302 crank has to be replaced by another 302 crank. Likewise with the 351W.

CONNECTING RODS

All 221, 260 and 289 engines use identical forged-steel connecting rods. *Center-to-center* length, as indicated in the chart is 5.155 in. This length is the distance between a connecting rod's wrist-pin and bearing-bore centers. These rods use 5/16-in. bolts and nuts to retain the caps. The standard 302 rod was shortened to 5.090 in. so the cylinder-block deck height could be maintained while increasing crankshaft stroke. The 302 retains the 5/16-in. bolts with one exception. A special heavy-duty rod was fitted with 3/8-in. bolts and nuts—C90Z-6200-B—but these are no longer available as replacement parts. So few were manufactured that you might as well forget them.

The Boss 302 rod is unique in that it retains the 5.155-in. center-to-center length of the previous small blocks. Deck-height clearance is maintained by reducing piston *compression height*—the distance from the center of the wrist-pin bore to the top of the piston. Boss 302 and HP289 connecting rods have something in common other than their center-to-center lengths. Both use 3/8-inch bolts. The one difference between them is how they are machined under the bolt heads. The HP289 rod is *broached* like the conventional rods whereas the Boss 302 rods are *spot-faced*. This makes the Boss rod slightly stronger because of less *stress concentration* on the inboard side of the bolt-head seat. So, if your are building an engine you plan to operate in excess of 6,000 RPM, consider using these rods. The HP289 rod—C3OZ-6200-C—has been replaced for service by the Boss rod—C9ZZ-6200-B. There are, however, two problems with using these rods. First, they cost over $50 *each*! This is too steep unless you're building an all-out racing engine. Then it's just steep. The other problem is they are rare and consequently are hard to find.

Ford high-performance parts are extremely hard to find and they are very expensive.

Not only was the 351W block and crankshaft changed extensively, so were its connecting rods. They were increased in length to 5.956-in. The *big-end* diameter is maintained, however the bearing inserts are thinner to accommodate the larger 2.311-in. crankshaft journals. Like its crankshaft, the only place you can use 351W rods is in a 351W block. The 351W rod also uses the 3/8-inch bolts.

PISTONS

Cast-aluminum pistons with internal steel reinforcing struts are used except in the Boss 302, which has forged-aluminum pistons. The biggest differences between the various pistons are dimensions to match a particular bore size and to be compatible with crankshaft stroke, block deck height and connecting-rod length.

A flat-top type piston with no *valve reliefs* was used in the 221 engine. Valve reliefs are cut-outs in the top of a piston to provide clearance for the valves when the piston is closest to the open valves near the top of its stroke. A *dome*, pop-up or raised portion on the top of a piston is used to increase an engine's compression ratio and a *dish* or cutout is to reduce it. The 221 piston had an extra groove, or *heat dam* above the top compression-ring groove to help shield the top compression ring from the hot combustion-chamber gases. This groove was dropped during the 1962 model year as it was found to be unnecessary.

Looking at the other pistons, the lower compression two-barrel-carburetor version of the 289 and all 302 and 2-V 351W use pistons with cutouts. Others use flat-top pistons with valve reliefs except for the 221 and Boss 302. The Boss 302 is unique because its top is domed *and* has valve reliefs.

A special word about the Boss 302 pistons. If you are rebuilding one of these engines and have to replace the pistons because of reboring, replace them with TRW forged pistons, L2324F or the later Ford service piston, D0ZZ-6108-A. Even though the production Boss 302 units were forged, their skirts had a cracking problem. This applies to the early service (replacement) pistons also. So, even if you don't have to install oversize pistons, check yours closely for cracked skirts. Replace any cracked pistons.

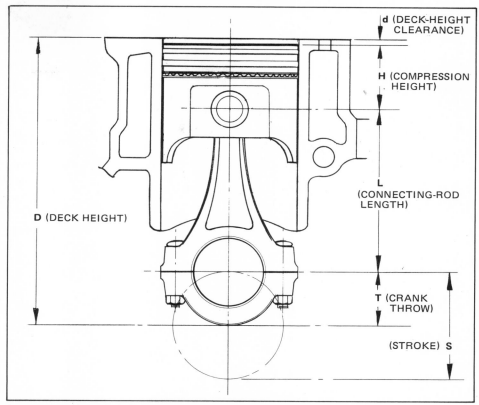

Crankshaft throw, connecting-rod length and piston compression height must add up to no more than a block's deck height, otherwise serious engine damage will result. When interchanging parts, use these figures to make sure there is sufficient deck-height clearance.

	221-289*	68-72 302	73-76 302	77-78 302	Boss 302	69-72 351W	73-76 351W	77-78 351W
D	8.206	8.206	8.229	8.206	8.209	9.480	9.503	9.503
d	0.016	0.016	0.034	0.0005	0.021	0.035	0.028	0.0145
H	1.600	1.600	1.605	1.616	1.529	1.739	1.769	1.782
L	5.155	5.090	5.090	5.090	5.155	5.956	5.956	5.956
T	1.435	1.500	1.500	1.500	1.500	1.750	1.750	1.750

Dimensions are in inches.
*Includes HP289.

Standard high-compression Boss 302 piston on the left and a standard low-compression piston on the right. Boss 302 piston has a dome or *pop up* to reduce clearance volume for increased compression—just as the '77 351W piston is dished to reduce it.

CRANKSHAFT, ROD AND PISTON MATCHING

What if you are replacing your connecting rods, pistons and crankshaft with those from another engine? Some interchanging between the 302 and the earlier small blocks is possible, but not the 351W because it has larger main-bearing journals. To do any interchanging, you must be aware of the basics. Otherwise, you can get into expensive trouble. For instance, crankshaft throw, connecting-rod center-to-center length, piston compression height and *deck-height clearance* must be compatible with the block's deck height. For a given stroke crankshaft, a longer rod will cause a given piston to operate *higher* in its bore. Installing a shorter connecting rod causes the same piston to operate *lower* in the same bore with the same crankshaft. This also applies to piston compression height. As compression height is increased—the wrist-pin bore is moved down in relation to the piston top—the top of the piston will operate higher in the cylinder bore. Reducing compression height moves it down. The last item, deck-height clearance, is simply how far short a piston at TDC is from coming even with a block's *deck* or gasket surface. This clearance is necessary to prevent a piston from hitting the cylinder head. The question now becomes, "How does all this relate to swapping parts?"

The *sum* of the crankshaft throw, connecting-rod length, piston compression height and deck-height clearance **must not exceed a block's deck height**, otherwise serious engine damage *will* result. This is particularly true if it is exceeded by more than the specified deck-height clearance. In this case the piston will come higher than the deck surface and impact against the cylinder head. On the other hand, if this sum is less, you'll have a very underpowered engine because its compression ratio will be low.

There are also other limitations to consider when swapping bottom-end components. A 302 crankshaft cannot be installed in a 221 even though the numbers add up right: the bottom of the 221 pistons will hit the connecting rods. 221 pistons are the only ones that can possibly work because of the 221s bore size. Oddly enough a similar situation exists with the 289 even though the 302 and 289 bores and piston compression heights are the same. 289 piston skirts will hit a 302s crankshaft counterbalances because 289 piston skirts are longer than a 302s. Consequently, 302 rods *and* pistons must be used when installing a 302 crank in a 289 block. This interchange of parts makes a 302 out of a 289. Since 1978, Ford service pistons for the 289-2V and '68-'72 302s are interchangeable.

One final and vital factor you must consider when changing bottom-end components is *balance*. Pistons, connecting rods, crankshaft *and* crankshaft damper and flywheel are matched for proper engine balance. Consequently, to ensure you'll end up with a balanced engine, particularly if the crankshaft and rods are changed, the complete assembly should be checked for balance. Many engine machine shops are equipped to do this.

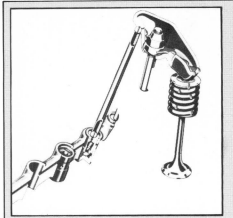

With some exceptions, small-block Fords have been characterized by their ball-type rocker-arm pivots. Photo courtesy Ford.

CYLINDER HEADS

Cylinder-head interchangeability gets very confusing, but not because the heads are so different—except for the Boss 302 heads which *are* radically different from all the others. Detail changes made to the heads over the years prevent interchanging heads in many instances. Another factor contributing to the cylinder-head-complexity problem is government emission standards and the design changes which have been made to meet these standards. All small-block heads will bolt on all small-block engines, but this does not mean they'll all work. Some will and some won't.

Cylinder-head Basics—To interchange cylinder heads you'll have to understand some cylinder-head basics. Let's start with compression ratio and how it is figured. As a cylinder's piston travels from BDC to TDC it *sweeps* through or displaces a volume called *swept volume* (S.V.) With the piston at TDC, the volume above it is called *clearance volume* (C.V.). Clearance volume includes the combustion-chamber volume in the cylinder head, volume created by the head gasket spacing the head above the block deck surface, additional volume created by piston deck-height clearance and the shape of the piston top. A concave or dished piston adds to the clearance volume. A convex or pop-up piston will reduce it. Of these four factors, cylinder-head combustion-chamber volume is the biggest contributor to an engine's clearance volume.

A cylinder's compression ratio is directly proportional to its clearance volume and its *swept volume* or *displacement*:

Compression Ratio
= $\frac{\text{Swept Volume}}{\text{Clearance Volume}} + 1$

ROCKER-ARM DESIGN

The ball rocker-arm fulcrum or pivot allows the rocker arms a higher *degree of freedom* than required to operate the valves. The rocker can pivot and rotate in all directions within limitations. To keep the rocker in contact with the valve Ford has used several different ways of restricting rocker movement.

Pushrod-guided—The original method stabilized the rocker arms by guiding the pushrods through close-tolerance rectangular holes in the cylinder head. These kept the rockers from moving sideways. The radiused tip of the rocker arm has a rolling movement against the valve tip as it operates the valve.

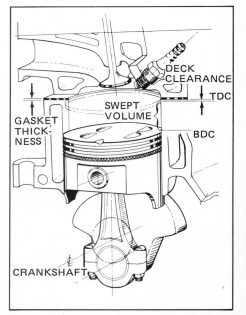

Compression ratio is determined by its swept volume and clearance volume. Swept volume is the volume an engine's piston displaces in its travel from TDC to BDC. It is also an engine's displacement divided by its number of pistons. Volume above a piston at TDC is clearance volume.

The compression ratio formula says that as an engine's *displacement is increased*, so must its clearance volume *increase to maintain the same compression ratio*. This is a very important consideration when interchanging cylinder heads, particularly these days when gasoline octane ratings have fallen like the value of the dollar. If you end up with a compression much over 9:1 you're going to have a difficult time finding fuel with a high enough octane rating to avoid detonation. If detonation is severe, engine damage will surely result. To see exactly what I mean, let's take a look at a couple of examples

Another style of pushrod-guided rocker arm uses separate slotted steel plates to control the pushrods and rocker arms. The plates are secured to the cylinder heads by clamping them under screw-in rocker-arm studs. Pushrod holes in the cylinder head are clearance holes which do no guiding of the pushrods.

Valve-stem-guided—Beginning with the hydraulic-lifter version of the 1966-1/2 289 and continuing in all small blocks through some 1978s, rail-type rocker arms are used. HP289s retain the early-style pushrod-controlled rocker arms. Rails cast into the end of the rocker arm straddle the valve tip. These rails restrict lateral rocker-arm movement by guiding against the tip of the valve stem, thus the designation *rail-type*. A clearance hole is drilled in the head casting for the pushrod. The load of stabilizing the rocker causes valve stem tips and rocker-arm tips to wear faster than is the case with pushrod-guided rocker arms. Valve guides are also loaded laterally, causing them to wear faster 90° to their normal wear.

As the rocker arm and valve-stem tip wears, the *rails* get closer to the valve-spring retainer. There is no problem until the rails begin to bear on the retainer. This condition can proceed to the point where the *keepers* release, letting the valve drop into its cylinder. Severe internal engine damage is the usual result.

Be aware of the danger of using worn valves and rail-type rocker arms in your rebuilt engine. If the rails are close to the spring retainers or are already contacting them, don't take chances—replace the rocker arms and valves.

Pivot-guided—Beginning in 1978, the rail-type rocker arm was phased out in favor of a pivot or fulcrum-guided rocker arm. This is the same type used in the Boss 302, Cleveland and "M" engines: 351C, 351M and 400M. The stamped-steel rocker arms pivot on pedestals. The pedestal is a stand and pivot in one assembly. The pivot is half a cylinder which is retained to the head with a bolt. The cylinder-type pivot restricts rocker-arm movement to the motion required to operate the valve. There is no stabilizing load on the valve, resulting in reduced valve-stem, rocker-arm and valve-guide wear.

All HP289s and small blocks prior to '65-1/2 used this close-tolerance hole (arrow) to guide the pushrod and control the rocker arm. The Boss 302 guide plate (arrow) accomplishes essentially the same thing.

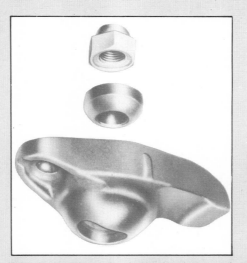 

Compared to the pushrod-guided rocker arm, the valve-guided or rail-type rocker arm is recognizable by rails straddling the valve-stem tip. This puts an additional load on the valve-stem tip and valve guide. Photos courtesy Ford.

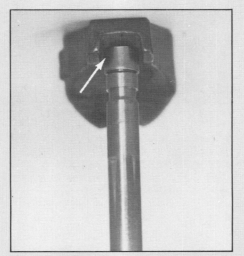 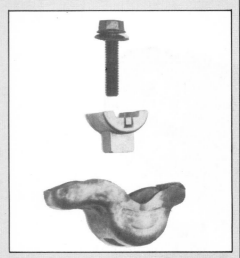

Valve-stem-tip caps (arrow) were introduced to combat rapid tip wear caused by rail-type rocker arms. Notice how the rails straddle the valve tip.

Rail-type rocker arm and ball pivot was phased out in 1978 in favor of this stamped-steel rocker arm with pedestal pivot.

from the compression-ratio chart. The 351W-2V, originally equipped with a 60.4cc (cubic centimeters) cylinder head, has an 8.7:1 compression ratio, and a 221 with its 45cc head also has an 8.7:1 compression ratio. Interchange heads on the two engines and they'll have compression ratios of 10.2:1 and 7.1:1, respectively. Put the 221 heads on the 351W-4V engine and the compression ratio goes to 13.2:1—guaranteed detonation! Reduced compression causes engine performance to suffer.

Another factor to consider when interchanging cylinder heads is valve size. Heads having valves too little for the engine won't *breathe* properly and power will be lost. In other words, the engine won't be able to draw in a sufficient fuel-air charge on the intake stroke and completely exhaust the burned charge on the exhaust stroke. On the other hand, if the valves are too big, particularly the intake valves, the velocity of the fuel/air charge entering the inlet ports will not be high enough to keep the fuel mixed with the air at low engine RPM. Lean mixtures and uneven mixture distribution will cause sluggish low-speed performance. Don't install heads with valves much more than 10 percent larger or smaller than the original valve size.

Other Considerations—When interchanging heads on '68 and later engines you have to consider whether they are to be used with an air pump. If your engine is air-pump equipped (Thermactor system) you'll have to install heads which will accept air-pump plumbing. Air-pump inlet ports must be plugged if you are using air-pump heads on an earlier engine without air injection. A special 1/2-20 plug, C6AZ-6052-A, is the Ford part for plugging externally manifolded cylinder heads. You'll need four per head. If these are not available, use cone-tipped socket-head type set screws with thread sealer.

When interchanging heads, pushrod lengths must be matched to the engine and the rocker-arm type used. Pushrod length is designed for each engine's deck height. 351W pushrods are too long for 221 through 302 engines and 221–302 pushrods are too short for the 351W. Positive-stop type rocker-arm studs and stamped-steel rocker arms require slightly longer pushrods than adjustable cast-iron rocker arms.

You may want to change your rail-type rocker-arm heads to earlier heads using the more durable pushrod-guided rocker arms. This is a practical swap on the later 289s and all 302s. I don't recommend it for a 351W because the smaller valves in the early heads will reduce performance and the smaller combustion chambers will push the compression ratio out of sight. Later 1978 heads with pivot-guided rocker arms will bolt right on in place of those using rail-type rocker arms.

 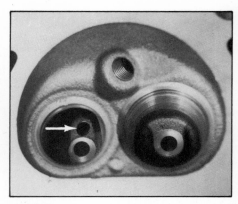

Large threaded hole (arrow) in the end of this '77 302/351W cylinder head is the inlet port for the internally-manifolded air-injection system. Fresh air is pumped into exhaust port through this hole (arrow).

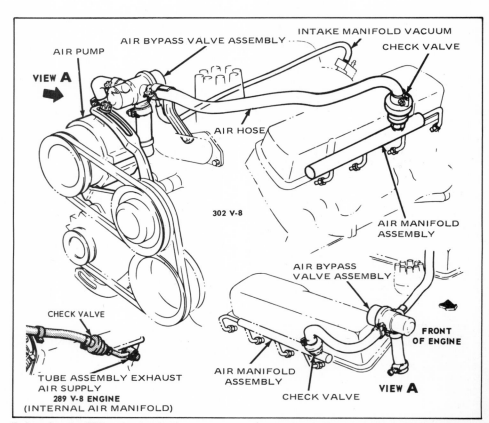

Except for the 289, early air-injection systems are externally manifolded like this 302 system at the top and right. 289 is internally manifolded with an inlet in the side of the head. All Boss 302 heads are internally manifolded as are all 302 and 351W heads from 1975 on. Drawing courtesy Ford.

Cylinder-head Evolution—Now that I've covered the ins and outs of interchanging cylinder heads, let's look at the evolution of these heads beginning with the 221 in 1962 and continuing through the 302 and the 351W.

221—1.59" and 1.39" intake and exhaust valves have 0.3094" and 0.3014" stem diameters. Rocker arms are the cast-iron adjustable type secured to the head with press-in studs and guided by the pushrods. These heads have the smallest combustion-chamber volume at 45cc.

260—Basically the same as 221 heads ex-

ENGINE	YEAR	STANDARD PUSHROD LENGTH (inches)
221–302	62–69½	6.825[1]
302	69½–77	6.905[3]
	77–78	6.883[2]
Boss 302	69	7.595
	70	7.660
351W	69–78½	8.170[3]
	78½–	8.205[2]

1. Before positive-stop rocker arms.
2. Used with stamped-steel rocker arms.
3. Oversize/undersize pushrods available in ±0.060-inch lengths.

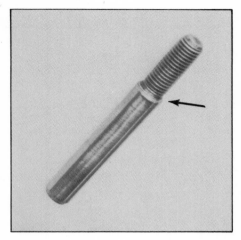

Positive-stop rocker-arm stud. Nut tightens down against shoulder (arrow) at the bottom of the threads fixing the rocker-arm and ball-stud positions relative to the cylinder head.

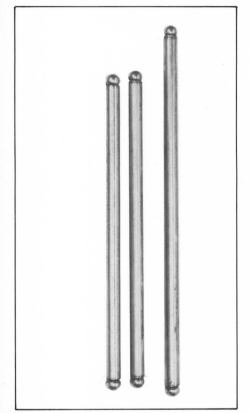

A comparison between the original 221-302 pushrods, slightly longer 302 pushrods used with positive-stop rocker-arm studs and the much longer 351W pushrod. Use the accompanying chart to determine standard pushrod length for your engine.

identical. 1966-1/2 marked the end of the pushrod-guided rocker arms. Rail-type rocker arms were used through 1968. In 1968, the last year of 289 production, 63cc combustion chambers made the 289 heads common with the new 302V engine.

HP289—This unique cylinder head has 49.2cc combustion chambers and 1.78" and 1.45" valves with 0.3420" and 0.3410" stems. Rocker-arm studs are threaded into the heads, giving the valve train a high-RPM capability. In 1964, combustion-chamber volume was increased to 54.5cc to make the basic castings common with the standard 289. Even though the standard 289 changed to rail-type rocker arms, pushrod-guided rocker arms were retained for the HP289.

302—Two basic cylinder heads were installed on 302 engines in 1968. One with 63cc combustion-chamber volume is common with the 1968 289. Another with 53.5cc chambers is installed on the 1968 high-compression 302-4V engines. These cylinder heads were machined two ways—with and without air-injection. Those machined for use with air pumps are externally manifolded for all years through 1974; internally manifolded from 1975 on. Beginning in mid-1968, non-adjustable positive-stop rocker-arm studs were installed and two-piece spring retainers were used. Hardened-steel caps on the exhaust-valve tips reduce tip wear caused by side-rail rocker arms. 1969 combustion chambers were reduced to 58.2cc.

302 and 351W heads are interchangeable starting with the 1977 model year with exception of the bolt-holes—351Ws are 1/16" larger so they'll accept 1/2" bolts. Combustion chambers are larger at 69cc. During 1978, the stamped-steel pivot-guided rocker arms gradually replaced the cast-iron rail-type rocker arms. Head castings are machined differently to accept pedestal rocker-arm pivots.

351W—Even though they are unique, 351W cylinder heads are basically the same as all other small-block Ford heads. Starting in 1969, combustion-chamber volume is 60.4cc. 1.84" and 1.54" valves with 0.3410" stems are used. Four extra intake-manifold bolts are accommodated by a corner cast into each of the two intake-manifold water passages. This changed the water-passage shape from a rectangle to an L shape requiring different intake-manifold gaskets. In 1975 the water passages became common with the 302s and the extra manifold bolts were dropped. 351W heads are internally manifolded for air injection from 1975 on. 351W heads became common with 302 heads in 1977 with the phase-in of stamped rocker arms in 1978. Combustion-chamber volume was increased to 69cc.

Boss 302—I talk about the Boss 302 heads last because they are totally different from all other small-block heads. They are more akin to 351C heads. Generally, Boss 302 heads are racing heads adapted to a street engine. Very large valves are angled in two planes for maximum breathing. Valve sizes are 2.23" and 1.71" in 1969; 2.19" and 1.71" in 1970. Stem diameter is o.3410". At 63cc for '69 and 58cc for '70, combustion chambers are large for a high-compression engine. This is due to their poly-angle shape. Shaped or pop-up pistons reduce some of the clearance volume. Boss 302 rocker arms are also unusual. They are stamped-steel, adjustable and pushrod-guided. The guides for guiding a pair of pushrods and rocker arms are plates secured to the head by the two screw-in rocker-arm studs. A loose stamped-steel seat is used between the spring and the cylinder head. Although only produced in 1969 and 1970, the Boss 302 engine was one of the best engines ever produced in terms of horsepower and durability potential for high performance. The cylinder heads were the key to this performance.

cept 54.5cc combustion chambers. Intake 1.67" and 1.45". Stems increased to 0.3420" and 0.3410", respectively.

289—1963 289 heads are the same as the 260, but with the 1.67" and 1.45" valves and 0.3420" and 0.3410" stems. The 260 shared this head in 1964-65. In February 1964, intake-valve diameter was increased to 1.78". To decrease compression ratio, dished pistons were used instead of changing the combustion-chamber volume. Beginning in 1966 some cylinder heads were machined to accept air-pump plumbing at each exhaust port, otherwise heads with or without air pumps are

MAJOR CYLINDER-HEAD CHANGES AFFECTING ENGINE PERFORMANCE AND PARTS INTERCHANGEABILITY

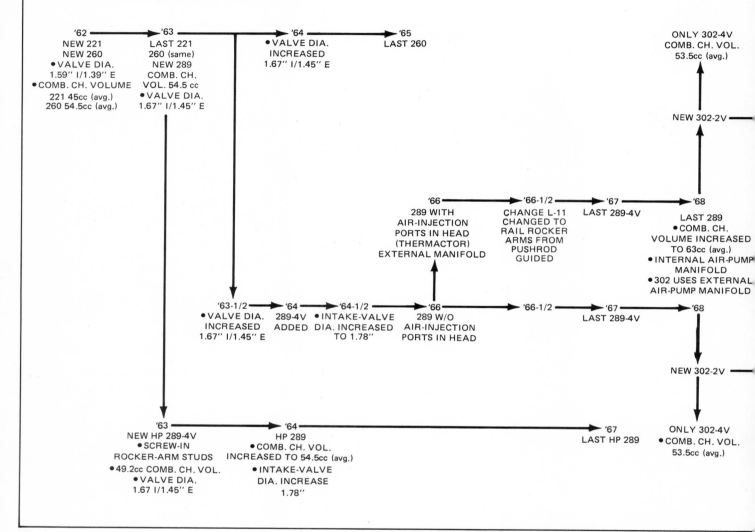

'62 → '63
NEW 221
NEW 260
- VALVE DIA. 1.59" I/1.39" E
- COMB. CH. VOLUME
  221 45cc (avg.)
  260 54.5cc (avg.)

'63
LAST 221
260 (same)
NEW 289
COMB. CH. VOL. 54.5 cc
- VALVE DIA. 1.67" I/1.45" E

'64 → '65
VALVE DIA. INCREASED
1.67" I/1.45" E
LAST 260

'66 → '66-1/2 → '67 → '68
289 WITH AIR-INJECTION PORTS IN HEAD (THERMACTOR) EXTERNAL MANIFOLD
CHANGE L-11 CHANGED TO RAIL ROCKER ARMS FROM PUSHROD GUIDED
LAST 289-4V
LAST 289
- COMB. CH. VOLUME INCREASED TO 63cc (avg.)
- INTERNAL AIR-PUMP MANIFOLD
- 302 USES EXTERNAL AIR-PUMP MANIFOLD

ONLY 302-4V COMB. CH. VOL. 53.5cc (avg.)
NEW 302-2V

'63-1/2 → '64 → '64-1/2 → '66 → '66-1/2 → '67 → '68
VALVE DIA. INCREASED 1.67" I/1.45" E
289-4V ADDED
INTAKE-VALVE DIA. INCREASED TO 1.78"
289 W/O AIR-INJECTION PORTS IN HEAD
LAST 289-4V

NEW 302-2V
ONLY 302-4V
- COMB. CH. VOL. 53.5cc (avg.)

'63 → '64 → '67
NEW HP 289-4V
- SCREW-IN ROCKER-ARM STUDS
- 49.2cc COMB. CH. VOL.
- VALVE DIA. 1.67 I/1.45" E

HP 289
- COMB. CH. VOL. INCREASED TO 54.5cc (avg.)
- INTAKE-VALVE DIA. INCREASE 1.78"

LAST HP 289

Shown here are five combustion chambers representing the basic small-block Ford designs: A. 54.5cc 260 combustion chamber is similar but larger than the 221's. B. 302-4V chamber is virtually identical to the 289's at a glance. C. More open 302-2V combustion chamber provides 9.5:1 compression as does D, the larger 351W chamber. Both have the same clearance-volume-to-swept-volume relationship. E. The later 301/351W has the largest small-block combustion chamber. Clearance volumes of the two engines are governed by the amount of piston dish and block deck height. Casting number D8OE-AB appears, but not engine displacement on this head. Notice the similarity between the 260 and 302/351W chambers—history is repeating itself.

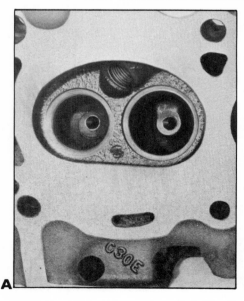

A

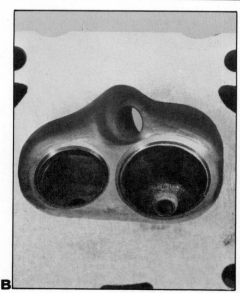

B

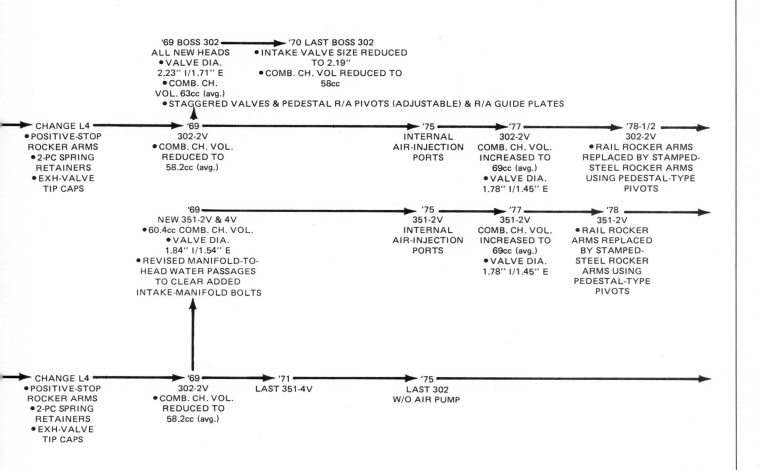

Chart illustrates mind-boggling evolution of the small-block Ford cylinder head. Much of the complexity is related to rocker-arm design changes and emission-standard compliance. During the years the changes have been subtle and usually confined to one area. For example, the L shaped 351W water passages were made common with the rectangular water passages of the 302 in 1975.

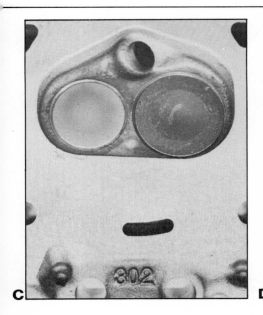

C

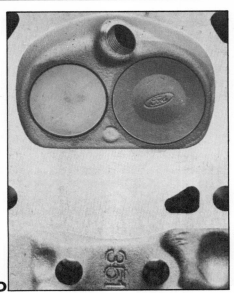

D

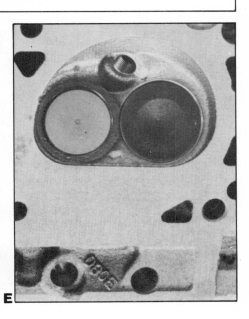

E

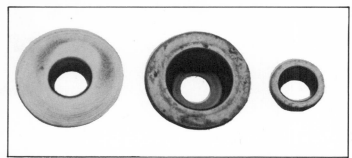

Two spring retainer styles you're likely to see are interchangeable. Single-piece machined one is on the left; two-piece stamped one on the right.

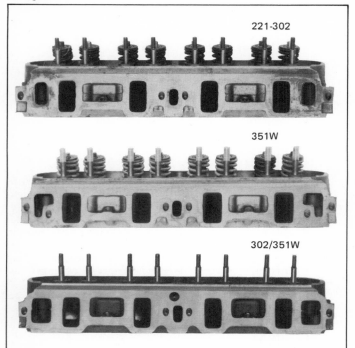

221-302

351W

302/351W

Comparison between 221—302, 351W and 302/351W cylinder heads points out differences in head-to-manifold water passages, number of manifold bolts and intake-port sizes. L-shape 351W water passages accommodate additional manifold bolts. Its intake ports are slightly larger than those in the other two heads. 221-302 and 302/351W water passages are rectangular in shape.

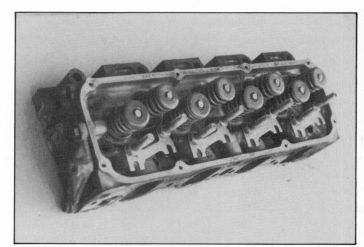

Boss 302 head is unique with very large ports and large canted valves which make it one of the best "breathing" cylinder heads in the world. It uses fully adjustable, stamped-steel and pushrod-guided rocker arms.

COMPRESSION RATIO/CYLINDER HEAD INTERCHANGE

COMBUSTION CHAMBER VOLUMES (cc's)	221	260	63-64 289-2V	65-67 289-2V	64 289-4V	65-67 289-4V	64 HP289	65-67 HP289	68-72 302-2V	302-4V	73-76 302-2V	77-78 302-2V	69 Boss 302	70 Boss 302	351W 4V	69-72 351W-2V	73-76 351W2-2V	77-78 351W-2V
45	**8.7**	9.9	10.2	10.6	10.1	11.5	12.3	12.3	11.4	11.3	9.2	12.2			13.2	11.4	10.2	11.4
54.5	7.6	**8.7**	**9.0**	**9.3**	**9.0**	**10.0**	10.5	**10.5**	10.0	9.9	8.3	10.5			11.5	10.1	9.2	10.1
49.2	8.2	9.3	9.6	10.0	9.6	10.8	**11.6**	11.6	10.7	10.6	8.8	11.4			12.4	10.8	9.7	10.8
53.5	7.7	8.8	9.1	9.4	9.1	10.2	10.7	10.7	10.1	**10.0**	8.4	10.7			11.7	10.2	9.3	10.2
63	6.9	7.9	8.2	**8.7**	8.2	9.0	9.4	9.4	**9.0**	8.9	7.6	9.4			10.4	9.2	8.5	9.2
58.2	7.3	8.3	8.6	8.9	8.6	9.5	10.0	10.0	9.5	9.4	**8.0**	10.0			11.0	9.7	8.9	9.7
60.4	7.1	8.1	8.4	8.7	8.4	9.2	9.8	9.8	9.3	9.2	7.8	9.8			**10.7**	**9.5**	**8.7**	9.5
*58													**11.0**	11.0				
*63													10.5	**10.5**				
69	6.4	7.4	7.7	7.9	7.7	8.4	8.8	8.8	8.4	8.3	7.2	**8.8**			9.7	8.7	8.0	**8.7**

*Interchanging Boss 302 cylinder heads is not practical. Requires the use of special pistons when used on other engines.

Compression ratios shown in bold face are for originally installed cylinder heads.

Never interchange cylinder heads without first checking the compression ratio. You could end with an engine that won't run on the available gasoline due to excessively high compression or one that performs like a dog because of low compression. This chart should help.

Teardown 4

1966 289-2V has all the characteristics of a good solid passenger-car engine: power, economy and durability. This engine is particularly suited to the Mustang and other vehicles in the 2800—3200-lb. weight class. Photo courtesy Ford.

Tearing down the engine is the first actual step in rebuilding. What you find through close inspection of your engine's components will "tell" you which parts have to be replaced or to what extent they must be reconditioned. It also gives you a first-hand look at the effects your servicing and the way you operated your car had on the internal wear. So don't look at the engine teardown as merely involving removing old parts and replacing them with new ones.

REMOVE EXTERNAL HARDWARE

While your engine is still "on the hook" after being lifted out of the engine compartment, remove as much external hardware as possible before setting it down. It will be more convenient to do it now, rather than after it's on the floor or a work bench. Another tip: Use small boxes or cans for storing fasteners and small parts. Store them in groups according to their function. For example, keep the exhaust-manifold bolts in the same container rather than in different boxes or cans containing some oil-pan, valve cover bolts, and the like. Be sure to drain the oil before setting the engine down.

Water Pump—Loosen the thermostat bypass hose clamp and then the mounting bolts. A *tap* on the end of the pump housing will break the pump loose from the timing-chain cover. Coolant will pour out of the two water passages when the seal is broken, so watch your feet!

Mark the Distributor Housing—Put a *scribe mark*, or scratch on the distributor housing and one on the engine block to match it, or simply remember the vacuum diaphragm points straight ahead. This provides a reference for positioning the distributor when you reinstall it. Disconnect the spark plug wires at the plugs and coil, and remove cap and wires. Remove the distributor after removing the hold-down clamp and disconnecting the vacuum-advance hose.

Exhaust Manifolds—Most engines have a sheet-metal duct for heating carburetor inlet air. It must come off before attempting to remove the right manifold. Disconnect the choke heat tube at the choke. Remove both manifolds, being careful with the right one so you don't break the choke heat tube, if there is one.

Engine Mounts—The engine mounts are still bolted to the block. Remove them, but keep track of the bolts by threading them into the block and tightening them.

Pressure Plate and Disc—Before removing your clutch, if your car has one, mark the pressure-plate cover and flywheel with a center punch so they can be replaced in the same relative position. When removing the clutch, loosen the 6 mounting bolts a couple turns at a time in rotation. This will prevent the pressure plate from being damaged as a result of being unevenly loaded. When the pressure plate appears to be loose you can remove the bolts all the way. Be prepared to handle about 25 pounds of pressure plate and disc. One final point, if your clutch or a portion of it can be reused, *avoid getting any grease on the friction surfaces,* particularly the disc. One greasy fingerprint can make a clutch chatter and grab.

Flywheel or Flexplate—Your engine has a flywheel if it was mated to a standard transmission and a flexplate if an automatic transmission was attached. There's no big difference about removing either except the flywheel weighs about 40 pounds so be ready for the weight. A 6-hole load-spreader ring is sandwiched between the mounting bolts and flexplate. Don't lose it. After removing the flywheel or flexplate, lift the engine plate from the back of the engine. It's the heavy flat sheet-metal piece which should be sitting loosely on the bellhousing or converter-housing dowel pins.

Carburetor—Remove the 4 carburetor-mounting nuts and lift off the carburetor and carburetor-spacer plate.

Drain Oil and Coolant—If you haven't drained the crankcase and the engine block coolant, do it now. It's the last chance you have to do it without creating an uncontrolled mess. I assume you won't

This tag has all the information on your engine: displacement, date produced and change level. Don't lose it!

Remove as much hardware from your engine as possible after pulling it out of the engine compartment and before setting it down. It will be light and easier to handle. The water pump, distributor, exhaust manifolds and carburetor are being removed here.

have any difficulties with draining the oil, otherwise you wouldn't be this far along. As far as the coolant which remains in the engine block, the easiest way of draining it is to knock out the *freeze plugs.* Most of the coolant will come out if you knock out the lowest plug on *each side* of the engine. Use a hammer and punch to drive the plug *into* the block. Watch out for the coolant because it'll come pouring out. To remove a plug from the block, pry it out with a set of Channel-lock type pliers and discard the plugs.

GETTING INSIDE THE ENGINE

Now's the time to prepare to open the "patient" up. Find a suitable place to work and set the engine down. A strong work bench with a work surface approximately 30 inches off the floor is ideal. However, if you're one of the few individuals who possesses an engine stand, by all means use it. If you don't have one, don't be concerned. I know professional rebuilders who prefer work benches!

INTAKE MANIFOLD

First remove the valve covers. Their edges hang over the inboard side of the cylinder heads and can interfere with the manifold's removal. Covers stuck to the heads can be popped loose by prying against the cylinder head and under the cover with a screwdriver after removing the attaching bolts. There's your first look at the inside of your engine.

After removing the intake-manifold bolts you'll quickly discover the valve covers were an easy touch compared to the manifold. It is stuck to both cylinder heads and the top cylinder-block surfaces. Wedge a screwdriver between one of the corners of the manifold and a cylinder head. Once the manifold comes loose, the rest is easy. When you're doing this, make sure you have *all* the bolts out. When the manifold breaks loose, lift it off—it's heavy! The valve lifters and pushrods will now be exposed.

REMOVING THE CYLINDER HEADS

Rocker Arms and Pushrods—Now it's time to zero in on the real serious stuff. Loosen the rocker-arm nuts or bolts so the rocker arms can be rotated to the side, freeing the pushrods so they can be lifted out. If your engine has rail-type rockers, don't remove the rocker arms until you're ready to disassemble the head. Inspection of the valve tips and their rocker-arm mating surfaces for wear can be done more easily. When you complete the rocker-arm and fulcrum removal, regardless of which type you have, wire the rockers and their fulcrums together in pairs. If they are mixed up you may end up with scored fulcrums and rockers when the engine is run. Rocker-arm stud nuts can go on in any sequence. To keep the pivots with their mating rockers the simplest method is to string them on a wire starting with cylinder 1 and proceeding to 8 in order of removal.

Unbolting the Heads—Each cylinder head is secured with 10 bolts—count them. Five are evenly spaced along the bottom of the head and 5 more are with the valve springs and valves. They are hidden from view when the valve cover is in place. I

When removing an automatic-transmission flexplate, don't lose the load-spreader ring (arrow). It must be reinstalled with the flexplate.

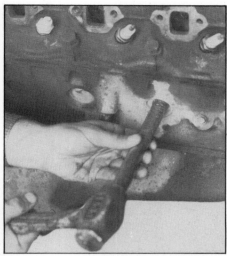 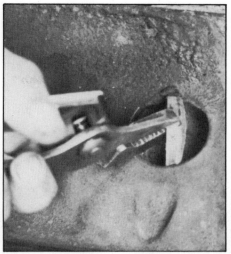

To remove freeze plugs, knock them in—then pry them out. Be ready for coolant to pour out of the block when the first plug is dislodged. The same occurs on the other cylinder bank.

may be insulting your intelligence by suggesting you make sure *all* the head bolts are removed before you try breaking the cylinder head loose from the block, but I've seen more than one enthusiastic individual attempting to pry a head loose from the cylinder block with a head bolt or two still in place. See the accompanying photo of a "two-piece" head that resulted because one fellow wouldn't give up. Don't let this happen to you. The last bolt to take out before you remove the right head attaches the oil dip-stick tube to the front of the head.

After my cylinder-head-bolt sermon, I'll now suggest you run a couple of bolts back in on each side just to engage a couple of threads. This prevents a head from coming off and falling to the floor. This may damage a cylinder head as well as your feet. After the heads are loose, remove the bolts and lift the heads off.

Extreme corrosion of these freeze plugs and rust buildup in the water pump was caused by using only water in the cooling system. Even if you live in the desert, use anti-freeze or rust inhibitor to prevent this happening to your engine.

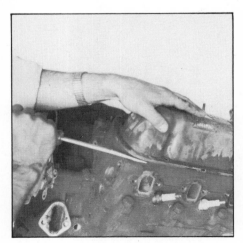 

Pop off your valve covers by prying under a flange at the corner. Note the new spark plug in cylinder 6. You'll see the reason for this later.

First look under the valve covers reveals extreme sludge buildup. This was caused by the engine being operated too cool because a thermostat was not installed. Always use a thermostat.

Cylinder heads can be as stubborn to remove as the manifold. After the bond is broken, the job is relatively easy. First, pry between the block and one of the top corners of the cylinder head. If that doesn't break it loose, wedge between the head and block. The wedge will have to be sharp like a chisel. If your engine is a 6-bolt-bellhousing 289 or 302, you have an easier job. Drive a chisel or similar wedge between the top-rear corner of the head and the bellhousing mounting boss. There is a small gap between the block and the head for getting the wedge started. Once the head has started to move, you can pry it the rest of the way and lift it off after removing the "safety bolts." Get both heads off and your engine will be 100 lbs. lighter.

 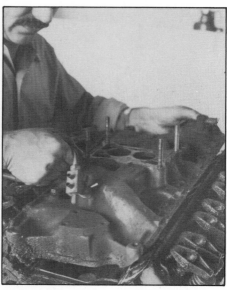 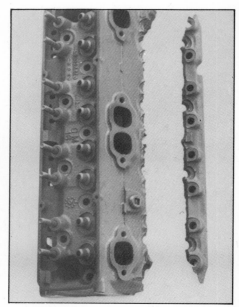

Your intake manifold should easily break free from the cylinder heads and the block by wedging a screwdriver between a corner of the manifold and a head. It should lift off easily once it's loose.

This is a tribute to human determination or stupidity, you be the judge. The bottom row of head bolts were not removed before the head was removed, but not all of it unfortunately. Although a brand-C cylinder head, the same thing can be accomplished with a Ford head with sufficient brawn, determination and a very long pry bar.

Begin the head-removal process by loosening up the rocker arms and removing the pushrods. If you elect to remove the rocker arms, keep the pivots and their rocker arms together—don't mix them up. Stringing them on a piece of wire is a simple and sure way of doing it.

Here's the reason for the odd plug in number-6 cylinder—oil. Compare it to the adjacent cylinders. Because of oil fouling a hotter plug was used in this one cylinder.

NOW FOR THE BOTTOM END

For the crankshaft damper removal, you'll need a puller. Before the front engine cover (timing-chain cover) can be removed, the damper pulley has to come off. The first thing you'll have to do is remove the center attaching bolt and washer. Just because the bolt is out doesn't mean the damper is going to slide right off. Quite the contrary, it must be removed with a puller. The damper fits on a long extension of the crankshaft. **Don't try prying the damper off.** The only thing you'll accomplish is damaging the damper and front cover or oil pan. You must use a puller with a plate matching the crankshaft pulley mounting holes in the damper and a center bolt that pushes against the crankshaft end. Because the threaded hole in the end of the crank can't be pushed against, insert the damper mounting bolt without its washer. This gives the puller something to push against and the bolt width clears the inside diameter of the damper as it is pulled off.

Roll the Engine Over—After the front pulley is off remove the oil pan. I do this because of the way it fits to the front cover. You'll need a 7/16-in. wrench for all the pan bolts except those next to the front and rear crankshaft seals are 1/2-in.

Oil Pump Assembly—Remove the oil pump. It's susceptible to damage now that it's exposed, particularly the pickup tube. After loosening the pump remove it and the drive shaft which inserts into the base of the distributor. The oil-pump drive shaft is one part you replace because they've been known to fail, resulting in seized bearings and other terrible things. Save it to compare length and hex size with the new one.

Front Engine Cover—Although front covers have changed since the 221 engine was first produced, their removal is similar with one slight exception. The 1968 through 1972 302s and the 1968 289 used a separate sheet-metal timing pointer attached to the cover by the lower-left cover bolt at one end and a self-tapping screw at the other end. The pointer hides one of the cover bolts. Attempt to remove the cover without discovering this and a broken cover may result. Swing the pointer or remove it to gain access to the "mystery" bolt. Loosen the self-tapping screw after removing the lower-left cover bolt and swing the pointer out of the way so you can get your wrench on the bolt. Leave the pointer on the cover so you don't lose it or its screw. With all the bolts out, remove the cover by giving it a light tap with a rubber or plastic mallet at the top-rear edge.

Small-Block Achilles Heel—Like most engines using a chain-driven camshaft, the chain and drive sprockets are a weak point when it comes to durability. This is especially true with the later nylon-jacketed aluminum camshaft sprocket which replaced the original cast-iron

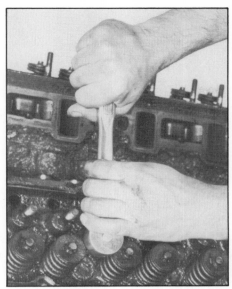

Remove all head bolts—there are 10—before attempting to remove a cylinder head. To break the heads loose from the block I wedge a chisel between the head and a bellhousing bolt boss. Don't wedge between the head and block deck surfaces or you'll surely damage them.

To remove the crankshaft damper you'll need a puller. Remove the damper attaching bolt and washer, then reinstall the bolt so the puller will have something to push against. The damper will clear the bolt head as it's being pulled off.

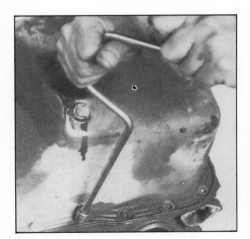

A speed handle comes in handy for removing the oil-pan bolts. I'm using a gasket scraper here to pop the pan loose from the block.

 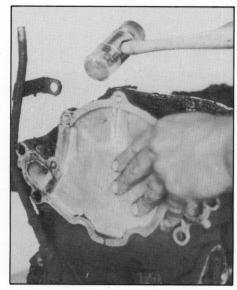

If your engine has a sheet-metal timing pointer attached to the front cover, swing it out of the way so you can remove the bolt hiding behind it. With all the front-cover bolts out you can remove the front cover. A light tap will break it loose.

With the oil pan out of the way, remove the oil pump and its driveshaft. Even though the driveshaft will be replaced save it for now, it'll come in handy later on.

This small-block finally refused to run after 165,000 miles because its timing chain wore and elongated to the point that it jumped several sprocket teeth, throwing the cam out of time with the engine.

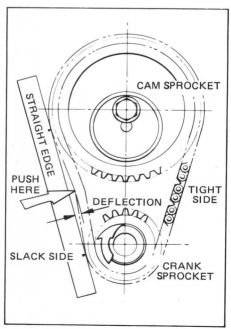

To check your chain, turn the crank to tighten one side of the chain and measure the slack side. If it deflects more than 1/2 in., replace it.

sprocket in mid-1965. The engine used for much of the photography in this book finally refused to run after more than 165,000 miles. Its timing chain was worn—elongated—so badly it jumped a few "cogs," so the cam was not timed with the rest of the engine. Consequently, the engine ceased to operate.

Prior to removing the chain and its sprockets, check them for wear if they've had less than 50,000 miles of use. If the chain and cam sprocket have been in the engine for more than 50,000 miles, don't bother checking—*replace them*. Check chain slack. It shouldn't exceed 1/2 in. Turn your crankshaft so one side of the chain is fully tight and the other side is slack. Lay a 12-in. steel rule or any straight edge against the chain as shown in the sketch and push in the middle of the unsupported length of the chain. With the straight edge for reference, your chain should have no more than 1/2-in. of slack, or deflection. If more, replace chain *and* the cam sprocket. If it has less, keep one point in mind. Your chain is going to start wearing the instant you get your engine back in operation, so the better the shape, the higher its useful mileage.

Getting the Chain Off—Remove the bolt and washer which hold the cam-drive sprocket and the fuel-pump cam to the camshaft. If you haven't already removed it, slide the oil slinger off the crankshaft extension. There's no slinger on 1977 and later engines.

To remove the timing chain and sprockets, you'll have to pull them part way off together. Start the cam sprocket moving by first prying against the backside of the sprocket on opposite sides with two screwdrivers. Work it loose. Don't use too much force, particularly with the aluminum/nylon type. Don't be surprised if the sprocket is a little stubborn, they usually are. This brings up a potential problem because the nylon teeth are easily damaged. It's not a problem if the sprocket needs replacing, but if it appears to be useable, try to save it. Spread the load against the back of the sprocket by placing a large-diameter flat washer between each screwdriver and the sprocket. The chain will also help distribute the

This is the fate of most high-mileage aluminum/nylon cam sprockets: cracked and broken teeth. Hot climates and emission controls which cause the engine to operate at higher temperatures aggravate the problem.

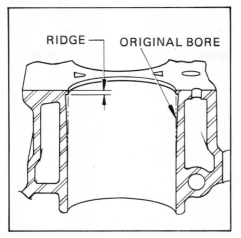
Slightly exaggerated, this is a section of a worn cylinder bore. A bore wears in a taper: more at the top than at the bottom. Short unworn section at the very top of a bore is the ridge, directly above the upper limit of travel of the top of compression ring.

A ridge reamer being used to cut away the ridge. This lets the piston and rod assembly slide up out of its bore without its rings hanging up and causing possible ring-land damage. If you intend to reuse your pistons, this is a must. Otherwise, it's a convenience.

load on the sprocket. If the nylon has become brittle from the heat, there's a good chance it will crack anyway. When you get the sprocket loose, move it forward off the end of the camshaft with the crank sprocket and chain. If you don't move them together, you won't be able to get the cam sprocket all the way off without the chain binding unless your chain is so far gone that you can slip it over one of the sprockets. Remove the cam sprocket and chain, and then the crankshaft sprocket.

If your engine used the aluminum/nylon cam sprocket, check the nylon for cracks and broken teeth. Even though the nylon-type gear wears very well, it gets brittle when subjected to high temperatures. Consequently, the teeth will crack and break off. So treat it like a chain. Replace it if it has more than 50,000 miles on it. Check it for cracks and breaks if it has less than this mileage and replace it if you find any. If you live in a hot climate, particularly if your engine is a '73 or later emission-controlled model which operates at higher temperatures, I suggest replacing the nylon sprocket with the earlier cast-iron type.

As for the cast-iron sprockets, they won't be cracked, but the plates which make up a *silent chain* may have worn ridges in the teeth to an extent that one or both of the sprockets should be replaced. To say exactly what the maximum allowable tooth wear is difficult. Drag your fingernail across the face of a tooth. If it is rough to the point of making your fingernail hang up or jump from ridge to ridge, replace the sprocket.

REMOVE THE CRANK, RODS & PISTONS BEFORE THE CAM

A camshaft can be removed before the crankshaft and rod-and-piston assemblies, but it's more convenient to reverse the procedure. With the crank and rods out of the way, you'll have access to the full length of the cam from inside the block rather than just the end of it from the outside. Again, it can and is done the other way, but it's more difficult and the crankshaft, rods and pistons have to come out anyway.

Remove the Ridge—Because the top piston ring doesn't travel all the way to the top of the cylinder when the piston comes up to its TDC, there is approximately 1/4-inch of unworn bore and carbon buildup at the top of the cylinder. This is called a *ridge*. It should be removed before you attempt to remove the rod and piston assemblies. This is especially true if your pistons can be saved because there is a good possibility they'll suffer ring-land damage if forced out over the ridge. If you *know* you aren't going to use your old pistons, they can be driven out the top of the cylinder and over the ridge using a long punch or bar which will reach up inside the piston. *Don't hammer on the rod.* Before doing this, just make sure you aren't planning on reusing these pistons because *they will be damaged!* A broken ring land ruins a piston.

To remove the ridges, you'll have to use a tool appropriately called a *ridge reamer*. This tool has a cutter mounted in a fixture which can be expanded to fit different bore sizes. A socket and handle from your tool chest rotate the tool to do the cutting, or reaming. As you use the tool, you'll have to rotate the crank a couple of times to move the pistons down the bores so you can get the tool in place. When cutting a ridge, cut the ridge *only to match the worn bore*. This is all you'll need to remove the piston, and any more could create a need for additional boring to clean up the cylinder.

Don't Mix the Connecting Rods and Caps—Now you can get on with removing the rods and pistons. Roll the engine over to expose the bottom end. **Before removing any connecting-rod caps**, make certain each rod and cap is numbered according to its respective cylinder. These numbers should appear on the small machined flat at the junction of each rod and cap. This check is a precautionary step because Ford stamps the numbers on during manufacturing. However, if they aren't numbered, you must do it because the rods and caps are machined as an assembly. Consequently, any caps interchanged with other rods will require re-machining before use.

If the rods and caps aren't marked, use a small set of numbered dies to do the marking *before* removing any bearing caps. Do the marking on the flats next to the bearing-cap parting line. If you can't obtain number dies, center punch *prick marks* on the rod and cap corresponding to the cylinder numbers. Or, use an electric engraver which has the advantage that you don't have to hit the rods or caps to mark them.

Be Careful of the Crankshaft Bearing Journals—Before removing a connecting rod and piston assembly, place something

45

Before removing a connecting-rod cap from its rod make sure both are stamped with the cylinder number. If they aren't numbered mark them on the machined flats at the rod and cap parting line like I'm doing here.

Turn the crank throw of the piston you're removing to BDC, then remove the rod cap. Place journal protectors (arrows) over the rod bolts and shove the piston and rod assembly out the top of its bore while watching that the rod doesn't hang up on the bottom of the cylinder.

over the rod bolts to protect the connecting-rod-bearing journals. If a bare rod bolt falls or drags against the bearing surface during piston and rod removal, the journal will surely be nicked. Therefore, place about a 2-in. piece of plastic hose or tube over each rod bolt to protect the bearing journals. Some parts manufacturers, suppliers and rebuilders have plastic sleeves which are flat with one closed end, especially for this purpose. They are available free of charge for the asking—all you have to do is find the right guy to ask. Your automotive supply store is a good place to begin.

To remove each piston, rotate the crankshaft so the piston is at BDC (bottom dead center). You'll have better access than if it were at TDC, and the rod will hang up between the crank throw and the bottom of the cylinder bore if it's in the middle of a stroke. Remove the nuts and bearing cap. With the protectors on the rod bolts, push the piston and rod assembly out the top of the cylinder. To protect your tender hands and knuckles, push on the piston or rod with the end of a hammer handle, using the hammer head as a handle. As you remove each rod and piston, put the bearings to one side. Then replace the cap on the rod and secure it with the nuts and set them aside.

Bearings Tell a Story—Don't throw your old connecting-rod or main bearings away yet—keep them in order. They'll save you the expense of farming much of your inspection work out because how a bearing wears is an accurate indication of the condition of parts such as the crankshaft and connecting rods. Therefore, as you remove each bearing set, tape them together with some masking tape and record which connecting rod or main cap the pair went with. It'll be valuable later on in the rebuild inspection and reconditioning stage.

With all the rods and pistons out of the way you can loosen the main-bearing bolts. The bearing caps should fit tightly in their registers so a little tap on the side with a hammer should pop them loose so they can be removed.

Remove the Crankshaft—After removing the main-bearing bolts, you'll find the main-bearing caps won't lift off like the connecting-rod caps did. The main-bearing caps fit tightly in registers machined in the bottom of the main-bearing webs of the block for accurate lateral (sideways) location. To remove them, tap the end of each cap while lifting up on it and it'll pop right off.

For 165,000 miles this crank looks excellent. A good cleaning and it'll be ready for another 165 thou.

Inspect the Bearings—Before tossing out the main bearings take a look at them; they'll tell you a story. Bearing inserts are made from plated copper-lead alloy or lead-based babbit, both on a steel backing or shell. If your engine has a lot of accessories such as power steering and air conditioning, the front top bearing should be worn more than the other bearings due to the higher vertical load imposed on the crankshaft and bearing by the drive belts. This wear is normal because of the way the crankshaft is loaded and should not be a cause for any concern. For the same reason, wear may also show up on the bottom of the center bearings, particularly at the second and third journals. If the bearings are copper-lead type, a copper color will show evenly through the tin plating. This makes it easy to distinguish wear because of the contrasting colors of the tin and copper. As for the lead-base bearings, it's more difficult to distinguish wear because of the similar colors of the bearing and backing.

You should be concerned about uneven wear from front-to-back on the total circumference of the bearing (top and bottom), scratches in the bearing surface and a wiped bearing surface. The first condition indicates the bearing journal is tapered, its diameter is not constant from one end to the other, causing uneven bearing and journal loading and uneven wear. Scratches in the bearing surface mean foreign material in the oil passed between the bearing and crankshaft journal. The usual cause of this is dirty oil and an oil filter which clogged, resulting in the filter being bypassed and the oil going through the engine unfiltered. A wiped bearing surface is usually caused by the journal not receiving adequate lubrication. This can be caused by periodic loss of oil pressure from a low oil level in the crankcase, a clogged oil passage or a malfunctioning oil pump. All these things have to be checked and remedied when the problem is found. Also, any problem you may find with the bearings means there may be damage to the crankshaft. Consequently you should pay particular attention to the crank journals which had bearing damage to see if there is corresponding damage to the crank.

Now, suppose you can remove a main cap without doing this and it's even loose sideways in the cap register. This indicates the cap or caps have been sprung or bent from being overloaded, usually from the engine *detonating*. Detonation is a condition where the fuel charge explodes rather than burning evenly. Detonation can be caused by the use of a lower octane-rated fuel than the engine requires or excessively high temperatures in the combustion chamber for one reason or another. It is commonly known as *pinging* or *knocking*. If any of your bearing caps appear to be sprung they can be reused in your rebuilt engine if it is intended for normal street and highway use. However, if you're going to use it for racing, the loose caps may have to be replaced. The block will then require line boring or honing to mate the bearing bores in the caps to those in the block and special bearings will have to be used.

When removing the main caps, don't remove the bearings yet. After you have all the caps off lift the crankshaft straight out of the block and set it beside the block. Place the bearing caps by their corresponding journals so you can see the bearing inserts. By doing this you'll be able to relate any bearing problems to the crankshaft journal each bearing was fitted to. Record the problem you discover with any bearing and the corresponding journal so you can take special care during your crankshaft inspection and cleanup, and possible reconditioning.

Removing the Cam—Two things are needed before you can pull the cam out. The most obvious is to remove the *thrust plate* which keeps the cam from moving forward. Remove the 2 retaining bolts, and then the plate, but don't attempt to remove the cam just yet. If you do, the result can be a damaged cam and lifters caused by the cam lobes and bearing journals running into the lifters as the cam is moved forward. You'll have to remove or raise the lifters out of their bores far enough so they'll clear the cam as it is being pulled out.

Varnish builds up on the lifter just below the lifter bore. Mike the diameter of the lifter at this point and you'll find its outside diameter is larger than the inside diameter of the lifter bore, making it difficult or impossible to remove through the bore.

The best way to remove the lifters is to remove the cam, then push the lifters out the bottom of their bores. To do this you'll have to push the lifters up into their bores and keep them there so the cam lobes and bearing journals will clear. *Keep* is the key word. Keep the lifters out of the way by turning the block over or standing it on its rear face. To move the lifters out of the way, turn the cam one revolution using the timing-chain-sprocket.

To start the cam moving forward, pry it with a screwdriver bearing against the back-side of a cam lobe and one of the bearing webs. Get ready to support the cam as it clear the bearings, particularly if

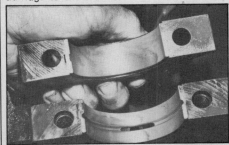

The main bearings look good too. This is the front one. Wear on the top bearing insert is due to the high vertical load imposed by the accessory-drive belts.

 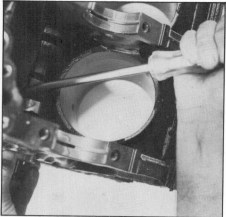 

Prior to removing your camshaft you should turn your engine block upside down and push the valve lifters away from their cam lobes up into their bores as far as they'll go. Remove the thrust plate and carefully pry the camshaft forward until it's free enough to move by hand. Slide it out the rest of the way, being careful not to bang the lobes into the bearings. Try to save the bearings as they wear very little and replacement requires special equipment.

These lifters were so heavily varnished they had to be driven out. No attempt was made to save them or the cam.

you are doing this with the block turned over. If you are intent on saving your cam bearings you should be doing this job with the engine standing up for the simple reason that gravity won't be working against you. A hard cam lobe can easily damage a soft cam bearing if the camshaft is dropped.

Keep the Lifters in Order—With the cam clear of the block, you'll be able to remove the lifters through the bottom of their bores. Before you start removing them, make some arrangement to keep them in order. If used lifters aren't installed in correct order so they mate with the original cam lobes, the mismatched lifters and camshaft lobes will be literally *wiped* out. You can use anything from a couple of egg cartons to a board with 16 one-inch diameter holes in two rows of 8. Just place them in the same order they would be viewed in as you look down into the lifter valley. Make sure you indicate the front of the engine on the device you are using to keep them in order.

Now all you have to do is push the lifters down through their bores. They'll go relatively easily because each lifter travels all the way down into its bore which prevents varnish buildup on the top part of the lifter. When pushing out the lifters, make sure you have one hand underneath ready to catch each one as it falls out to prevent damage. On the other hand, if you don't plan to reuse your cam you can just push them out because you won't be reusing them anyway.

Old and new lifters. Old lifter is useless as it's worn from its new convex spherical shape to a concave shape. The camshaft is also worn beyond reuse as shown by the wear pattern (arrow) across the full width of the lobe.

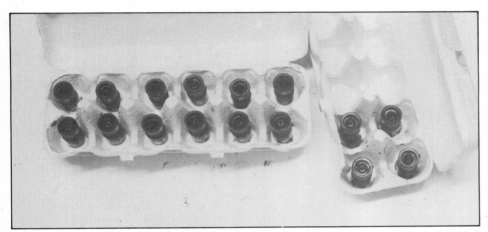

If your camshaft and lifters are salvagable you *must* keep the lifters in order for reinstallation in their original bores. I'm using a couple of egg cartons here.

Plugs, Cam Bearings and Things—After the removal of the lifters, your engine block should be free of its moving parts. What's left are the oil-gallery plugs, camshaft bearings and rear plug, the oil filter adapter and the water-jacket freeze plugs—*if* you didn't remove the freeze plugs as suggested earlier in the chapter.

Oil-Gallery Plugs—There are 6 oil-gallery plugs, 3 at the front of the engine and 3 at the rear. The 3 at the front are *soft plugs* and the 3 at the rear are threaded into the block. If your engine is fairly clean inside you needn't remove these threaded plugs because they can be difficult and it's probably unnecessary. However, if your engine is sludged up and fairly dirty, or you'll feel better if you do it— here's how.

The three plugs at the rear are usually *very* tight. Removing them is difficult

To clean the oil passages, remove the front and rear oil-gallery plugs. Remove the front cup-type plugs by drilling a 1/8-in. hole in the center, then thread a sheet-metal screw into the plug. Now you have something to get hold of. Clamp onto the screw head with some Vise-grip® pliers and pry it out with a screwdriver.

If your cam bearings must be removed, here is one method of doing it—using a mandrel and drive bar. Mandrel must fit the bearing ID, shoulder against the bearing and clear the bearing bore in the block as the bearing is being removed. At least three mandrels are required because of the different bearing sizes.

without rounding the *square* off first. Use a tool which will minimize this possibility—such as a square socket. Chances are you don't have one of these, so some fairly large Vise-Grips® are the next best bet. They'll mess up the square somewhat but you'll get the plug out. The front plugs can be easily removed by driving them out from the backside with a long 1/4-in. rod. If you don't have one handy or haven't removed the threaded plugs, the next approach is to drill a small hole in each plug—1/8 in. will do. Thread a sheet-metal screw into the hole and clamp onto the screw head with a pair of Vise-Grip® pliers. Pry against the plier jaws with a screwdriver until the plug comes out.

Camshaft Bearings—Replacement cam bearings are precision inserts which assume the correct diameter after they are driven into the block. Now, because the bearing bores in the block are bored in line, or line-bored, replacement bearings installed using conventional methods should yield the same alignment results as the original factory method.

I am explaining cam bearing installation now so you'll understand the importance of avoiding damage during teardown and cleaning if you intend to reuse them. Be aware that a hot-tank solution, so ideal for cleaning cast iron and steel, dissolves aluminum or babbit. However, the bearings can be saved *and* the block cleaned if a spray-jet tank is used, *if the cleaning solution is not caustic.*

Remember, camshaft bearings wear so little that the normal set could outlast two or three engines. With these points in mind, you'll have to decide now whether you're going to attempt to save the bearings or not.

Removing the Cam Bearings—If you are entertaining thoughts about removing the cam bearings yourself, don't! Unless you have an engine machine shop or access to camshaft bearing installation and removal equipment and know-how to use it, farm this one out.

Cam-bearing tools vary in sophistication from a solid *mandrel* which fits inside of *its* bearing—you need five mandrels—to a more sophisticated *collet* mandrel which expands to fit the bearing ID. Both types of mandrels have a shoulder which bears evenly against the bearing shell for removing or installing a bearing. The bearings are removed by a drive bar which centers in the mandrel, or by a long threaded rod which pulls the mandrel.

Either type is fine for bearing removal, but the latter is the best for installation because it ensures the bearing will be square in its housing and to the cam centerline. The chance of bearing damage is minimal, particularly for someone who doesn't do engine rebuilding on a regular basis. I'll take up bearing installation when it's time to do so. If you have the equipment and have elected to remove your cam bearings, do it now. Remove the rear cam plug, then the bearings. Use the same method you used on the front oil-gallery plugs, or drive the plug from the inside of the block.

Oil-Filter Adapter—With the removal of the oil-filter adapter, your engine block will be as bare as it's going to be with the exception of the grease, dirt and old gaskets. All you'll need for removing the adapter is a *big* socket—1-1/4-in. to be exact. Just be careful when attempting to break the adapter loose. The chamfer inside the socket will only allow a small part of the hex to engage the socket flats. Consequently, if the socket slips once, the hex will round off and you'll have an increasingly difficult time getting it off. So to break the adapter loose, push on the socket to prevent the socket from slipping off the hex. If you are going to do a lot of work on these engines, grind the socket end to eliminate the chamfer, thereby making adapter removal super easy.

With the removal of the oil-filter adapter, your engine block should be completely bare.

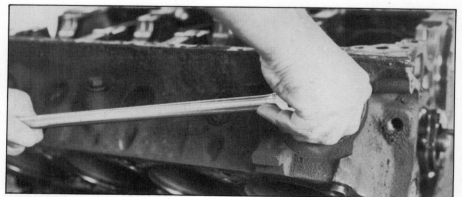

When removing the oil-filter adapter, shove against the socket to keep it from slipping and rounding off the corners of the hex.

Inspecting and Reconditioning the Shortblock 5

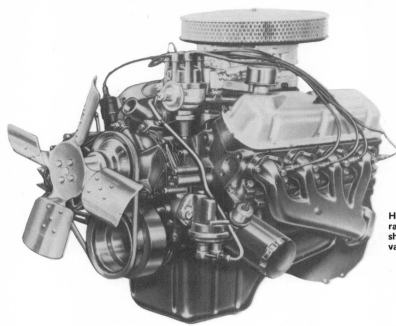

HP289 provides fierce competition in professional and amateur racing, yet it remains a practical street engine. Vacuum can shown on the distributor is not correct. HP289 did not use vacuum advance. Photo courtesy Ford.

Most people have an inherent knack when it comes to taking things apart, so I'll assume you've had no real problems to this point. Problems usually occur during the inspection, reconditioning and assembly processes, but don't show up until the engine is back in the vehicle and running—or not running. More often than not, goofs are due to insufficient information which translates into a lack of knowledge. I've tried to include all the information you will need. Special attention has been paid toward including information which is especially hard to get anywhere else. Therefore, if you apply this information with an abundance of common sense and a reasonable amount of care, your rebuilt engine should perform better than when it was new.

CYLINDER-BLOCK CLEANING AND INSPECTION

One of the most important jobs is to clean each component that will go back into your engine. After you've cleaned everything so thoroughly you are sick of it, you'll have to *keep all of it clean.*

Getting It Clean—If for no other reason, the size and complexity of an engine block makes it the most difficult component to clean. The job must be done right because the block is the basic foundation which is responsible for supporting, cooling and lubricating all the other components. Consequently, if the cleaning job is short-circuited, one or all of these functions will be compromised, so *get it clean!*

Before destroying the evidence, inspect the head gaskets for leaks. They will show up as rust streaks on the cylinder-head or cylinder-block mating surfaces if it is a coolant leak, or black or gray streaks radiating from a combustion chamber in the case of a more harmful compression leak. If you find a compression leak connecting a combustion chamber with a water passage, you probably experienced trouble with coolant loss due to the coolant system being over-pressurized. On the other hand, a compression leak vented to the atmosphere or another cylinder will have shown up only if you performed a compression check on the engine *prior* to tearing it down. If you discover a leak, check the block and head gasket surfaces for warpage or other imperfections after you have them well cleaned. One type of leak to be particularly watchful for is the type connecting two cylinders. Check the block for *notching* at this point. Notching occurs when hot exhaust gases remove metal much in the same manner as an acetylene torch. Notching is more prevalent with racing engines, but it also happens when any engine is driven a long time before a blown gasket is fixed. Check the severity of any notches after cleaning the block's deck (head-gasket) surface.

Scrape The Gaskets Off—Get the worst job out of the way first by scraping all the gasket-sealing surfaces. A gasket scraper is a tool specifically designed for the job. Now's the time to invest in one. If you try doing the job with a putty knife or a screwdriver and then switch to a gasket scraper, you'll find out what I mean. You could've saved the skinned knuckles, bad temper and loads of time. Surfaces which will need scraping are the two cylinder decks—which are the meanest—the front-cover, oil-pan, intake-manifold and oil-pump gasket surfaces. Don't stop with the block. While you're at it, do the heads and intake manifold too.

Boil It, Spray It or Hose It Off—The cleaning process can be handled several ways. Probably the best method is to truck the block to your local engine rebuilder for *hot tanking*. This process involves boiling the block for several hours in a solution of caustic soda—the longer it's in, the cleaner it'll be. Don't just have the block done, have the crankshaft, heads and all the other *cast-iron and steel* parts done at the same time. Throw in the nuts, bolts and washers too, but *don't include* any aluminum, pot metal, plastic or materials you'll want to see again because, like the cam bearings, the hot tank will dissolve them! Now, if you are determined to retain your cam bearings, you'll have to find a rebuilder who uses a *non-caustic* spray degreaser. Volkswagen engine rebuilders often use this type of cleaner because of the wide use of aluminum in VW engines. Other techniques you can employ are steaming, engine cleaner and the local car wash, or the garden hose with cleaner, detergent and a scrub brush. Regardless of which method you use, concentrate on the engine's interior, particularly the oil galleries and holes. Use a rag with a wire to drag it through the galleries. Resort to round brushes, pipe cleaners or whatever, but pay particular attention to the oiling system. Many supermarkets carry nylon coffee percolator brushes which are suited to this job. Gun brushes work well too. Team these up with Ford's Carburetor and Combustion Chamber Cleaner and you'll be able to do an excellent cleaning job.

Don't Forget the Threads—After you think you've gotten the block as clean as it's going to get, *chase* the threaded holes

51

Someone has to do it. Make sure all the gasket surfaces are scraped clean of old gasket material.

Why is Len smiling? This block is being hot-tanked. Don't have your block cleaned by this method if you intend to save your cam bearings or they'll have to be replaced for sure.

Pay particular attention to the oil galleries when cleaning your cylinder block. Steam does the best job.

in the block. All that's involved in chasing a thread is running a tap all the way in the thread to clean it. A *bottom* tap, as opposed to a taper tap, which is used to start a thread, is the best type to use. You'll be shocked at the amount of crud the tap will extract from the threads, particularly after the "spic-and-span" job you gave the block. This procedure is especially important for the bolts which must be torqued during assembly such as the head-bolt and main-bearing-cap-bolt threads. The sizes are 1/2-13 (1/2-in. diameter, 13 threads per in.) for all small blocks except the 351 which uses a 9/16-12 thread.

Now go after the water passages. Remove any loose rust, deposits and core sand. Pay particular attention to the passages which connect the cylinder heads to the block to ensure a good flow of water between the block and heads. A round, or rat-tail file works well for this job, but be careful of the cylinder-head gasket surfaces. A deep gouge can be responsible for a head-gasket leak. Give the same treatment to the cylinder heads.

During all this scrubbing, scraping and general clean-up, a source of compressed air for forcing dirt out of hard-to-get-at areas and drying the block of moisture will be a definite help. Controlling moisture becomes more of a problem as you recondition more and more of the various engine parts. Bearing surfaces, cylinder bores, valve seats and any machined surface will rust just from the humidity in the air. Prevent this by coating the machined surfaces with oil after cleaning. Several brands of spray-on oil are available at your local store, such as WD-40® or CRC®. They'll do the job with a lot less fuss and mess than trying to use a squirt can of motor oil. Whatever you use, don't leave any freshly machined surfaces unoiled or they will rust for sure.

CYLINDER-BLOCK FINAL INSPECTION AND RECONDITIONING

Inspecting the engine block to determine what must be done to put it back in tip-top condition is your first reconditioning step. To perform a satisfactory job of inspection you'll need 3- to 4-in. inside and outside micrometers, a *very* straight edge and some feeler gauges. You'll also need a 2- to 3-in. outside mike for checking your crankshaft main- and rod-bearing journals. You may not need the straight edge if the head-gasket seal checked OK. If the old gasket didn't leak, the new one won't either if it's installed correctly.

Checking Bore Wear—Cylinder-bore wear dictates whether they need boring or just honing. This will, in turn, largely determine if you have to install new pistons—no small investment.

You can use three methods to check bore wear. The best is a dial-bore gauge, but you probably won't have access to one of these, so let's look at the remaining methods. Next in the order of accuracy is the inside micrometer or telescope gauge and an outside mike. The last method involves using a piston ring and feeler gauges to compare end-gaps at different positions in a bore. These will correlate to the bore's *taper*.

Bore Taper—Cylinder walls don't wear the same from top to bottom. A bore wears more at its top with decreasing wear down the bore. There's virtually no wear at the bottom of a cylinder. More pressure is exerted on the cylinder walls by the compression rings at the top of the stroke, decreasing through the length of the stroke as the piston travels down the cylinder. The bottom of the cylinder is well lubricated, stabilizes the piston and receives little wear. This is shown by the upper part of the bores which are shiny—while the bottom retains its original *cross-hatch* or hone marks put there in the factory. A vertical section through the cylinder wall shows this taper when compared to a section through an unworn cylinder.

Measuring Taper—Because a bore wears little at the bottom, if you compare the distance across the bore at the bottom versus that at the top, just below the ridge, you'll be able to determine its taper. Also, bores don't wear evenly all the way around, nor do all cylinders wear the same. Therefore, when measuring a bore, measure it parallel to the centerline of the engine first, and then 90° to the centerline. Take a couple of measurements in between and use the highest figure. It will determine what and how much will have to be done to that cylinder to restore it. Also, because it's not practical to treat each cylinder separately, you'll have to pinpoint the one with the worst taper and let it be the gauge of what must be done to the remaining seven cylinders. One exception to this rule is when one cylinder is damaged or worn beyond the point where it can't be restored by boring and the other cylinders are OK. In this case, it might be less expensive to have the cylinder *sleeved* rather than junking the block and buying another one.

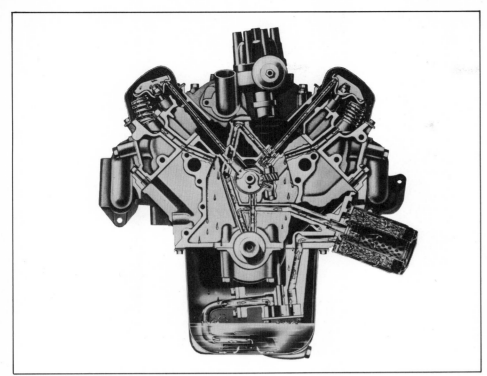

Lubrication system of the small-block-Ford engine. Photo courtesy Ford.

to its original centerline usually requires removing more metal than indicated by the taper figure. As a result, final bore-size determination is made at the time of boring. If a cylinder bore does not clean up at 0.010 in. oversize, the machinist has to bore to 0.020 in., the next available oversize. He starts boring the cylinder with the worst wear/taper because this should establish the maximum bore for all of the cylinders.

Dial-Bore Gauge or Micrometer Method— When using a bore gauge or mikes to

You'll notice when checking bore wear that the end cylinders are worn the most. The reason is the cooler a cylinder operates, the more it wears. This wear is concentrated at the portion of the end cylinder wall closest to the ends of the cylinder block in cylinders 1, 4, 5 and 8. A quick way to verify what I've said is to compare the ridges of each cylinder by feeling them with your finger tip, particularly the variation in thickness of the ridges in the end cylinders. Because of the wear pattern in the end cylinders, wear measured parallel to engine centerline will exceed that measured 90° to the centerline.

Due to uneven cylinder bore wear, taper, or the difference between the worn and unworn portion of a bore, measuring will not provide a final figure of how much a cylinder must be bored to clean it up, or completely expose new metal for the full length of a bore. The reason is uneven wear shifts a bore's centerline in the direction of maximum wear, giving it an irregular shape. To restore a cylinder

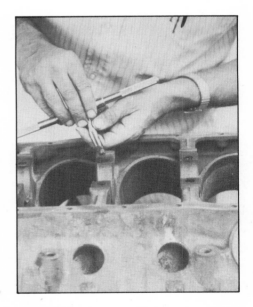

Pay particular attention to details. Here I'm using a tap to chase the main-bearing threads and a rat-tail file to remove rust from the water-passage openings in the block.

Why you should always use a thermostat. Bore wear increases dramatically as an engine is operated cooler than 180°F (82°C). Data courtesy Continental Motors.

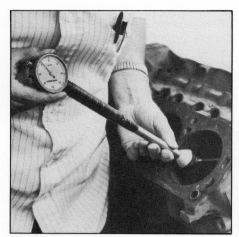

Checking bore taper with a dial-bore gauge. Bore wear is read directly.

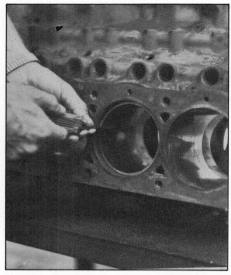

$G_2 - G_1$ ΔG	TAPER
0.000	0.0000
0.001	0.0003
0.005	0.0016
0.010	0.0032
0.015	0.0048
0.020	0.0064
0.025	0.0080
0.030	0.0095
0.035	0.0111
0.040	0.0127
0.045	0.0143
0.050	0.0159

Approximate Taper = $0.30 \times \Delta G$

Using ring end-gap to determine bore taper. Maximum ring gap is found with the ring placed immediately below the ridge. Minimum gap is measured with ring pushed down in the unworn section of the bore. Use a piston inserted upside-down in the bore to square the ring up before checking gap. After finding the maximum difference between two end-gap readings, use this curve or chart to determine bore wear. ΔG is the difference between G_2 and G_1.

measure taper, measure the point of maximum wear immediately below the ridge. Because the wear will be irregular, take several measurements around the bore to determine maximum wear. To determine taper, subtract the figure at the bottom of the bore from the maximum bore at the top.

Ring-and-Feeler-Gauge Method—Measuring bore wear with a ring and feeler gauge is going about it indirectly. When using this method, you actually compare the difference between the circumferences of the worn and the unworn bore. The accuracy of this method is less the more irregularly a cylinder is worn, however it is accurate enough to determine if you'll need to bore and fit oversize pistons or clean your pistons, hone the bores and install *moly rings*.

To use the ring-and-gauge method, place a ring in the cylinder and compare the difference in ring gaps with the ring at the bottom of the bore and with it at the top of the bore—immediately below the ridge. Use the same ring. The ring must be square in the bore to get an accurate reading. To square the ring, push the ring down the cylinder with a bare piston—no rings—to where you want to check ring gap. Do this with your feeler gauges and record the results. After measuring the ring-gap difference, use the accompanying chart or graph to determine bore taper. Note that *taper is approximately 0.3 times ring end-gap difference*.

How Much Taper Is Permissible?—To answer this question, ask yourself how many more *good* miles you want out of your engine. Do you want it to last 10,000, 20,000, or 100,000 more miles before the oil consumption takes off? If your engine has excessive taper, even new rings will quickly fatigue and quit sealing. They must expand and contract every stroke of the piston to conform to the

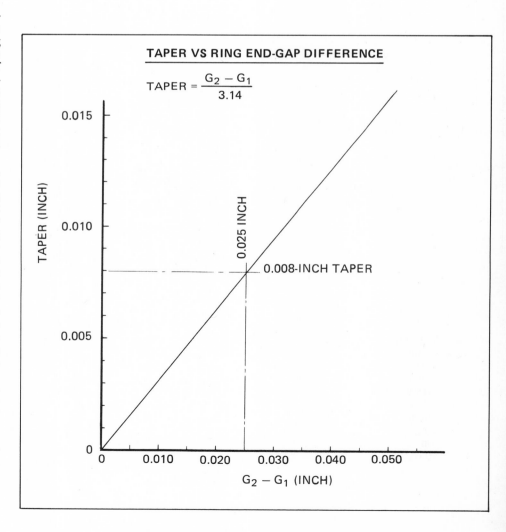

TAPER VS RING END-GAP DIFFERENCE

$$\text{TAPER} = \frac{G_2 - G_1}{3.14}$$

0.025 INCH
0.008-INCH TAPER

SLEEVING AN ENGINE

A sleeve is basically a portable engine cylinder. It is used to replace or restore a cracked, scored or otherwise damaged bore that can't be restored by conventional boring and honing techniques. An expenditure of only $20 to $30 for sleeving a cylinder can save you the cost of a new or used engine block. A sleeve is a cast-iron cylinder which is slightly longer than the length of the cylinders in the engine it is made for. It has a smaller ID than the original bore for finishing stock and a wall thickness varying from 3/32 to 1/8 in.

To install a sleeve, the damaged cylinder must first be bored to a size 0.001" less than the OD of the sleeve. Rather than boring all the way to the bottom of the bore, the boring machine is stopped just short to leave a *step*, or *shoulder* for locating the sleeve. The step and the interference fit of the sleeve prevent it from moving around after the engine is back in operation. Some rebuilders leave more than 0.001" interference between the sleeve and the block. The problem is this overstresses the block and can distort neighboring cylinders, consequently the 0.001" figure should be adhered to.

When an engine is ready to be sleeved, it should be evenly warmed up using a torch, furnace or whatever so the block will *grow*. At the same time the sleeve is cooled to shrink it—a refrigerator freezer works great for this. The sleeve will *almost* drop all the way into the cylinder, however it will quickly rise to the same temperature as the block, assuming its interference fit. Then it must be pressed or driven in the rest of the way and sealed at the bottom. Some shops don't bother with heating and cooling and drive in the sleeves in all the way. The excess length of the sleeve protruding from the block is trimmed flush with the deck surface. The block is now restored and the sleeved cylinder is bored to match other cylinders.

Checking piston-skirt-to-bore clearance. If it's 0.008 in. or more you should rebore.

Embossed pattern (arrow) rolled into this piston's skirt is a knurl. It's a temporary fix for taking up clearance between a worn bore and piston. Photo by David Vizard.

varying bore. Consequently, they lose their resiliency, or springiness. Also, because ring gap must be the correct minimum dimension at the bottom of the bore, if the engine has taper, the gap will be larger at the top where compression pressures are the highest. The results are reduced combustion pressure, lost power and increased blow-by.

Therefore, if you just want a "band-aid" job on your engine so it will go another 10,000, maybe 20,000 miles before you're right back where you started, you can *get away* with rerining a block which has some cylinders measuring as much as 0.010" taper. If you are using the ring-and-feeler method, rebore if taper exceeds 0.008". This is because this method masks some forms of wear. However, if your object is to have a *truly* rebuilt engine, don't merely rering if taper exceeds 0.006" using the dial-bore gauge or micrometer methods of checking, or 0.004" by the ring-and-feeler method. Remember, taper gets worse with use, not better. It's best to start a rebuilt engine's life with a *straight* bore and new pistons if you're after maximum longevity.

Piston-to-Bore Clearance—All this talk about whether to bore or not may turn out to be purely academic. The main reason for not wanting to bore an engine is to avoid the cost of new pistons, a legitimate reason, considering they cost $10 to $20 each. If your old pistons are damaged or worn to the point of being unusable, you'll have to purchase new pistons whether your block needs boring or not. So, to make the final determination as to whether your engine should have a rebore, check the piston-to-bore clearances. Refer to the piston section of this chapter for other piston-related problems you must check for before reusing your old pistons.

Two methods can be used for checking piston-to-bore clearance. The first is mathematical. Measure across the cylinder bores 90° to the connecting-rod wrist pin to determine maximum bore diameter in this direction. Then, measure the piston *which goes with the bore*. Mike it in the plane of the wrist pin and 90° to the wrist-pin axis, measuring the piston across its thrust faces. Subtract piston diameter from bore diameter to get piston-to-bore clearance. Measure with the block and piston at the same temperature. If the temperatures are too far off, your measurements will not be accurate.

Directly measuring the clearance is the second method. Install the piston-and-rod assembly in *its* bore less rings, and in its *normal* position. By normal position, I mean the connecting-rod wrist pin is pointing toward the front of the engine. This will ensure that the clearance you're checking is what the piston *sees* when it is correctly installed in the engine. Use your feeler gauges from the top of the cylinder to check between the piston's thrust face and the bore in several positions up and down the bore between the piston's TDC and BDC. Again, the maximum clearance should not exceed 0.008 inch.

There is a method to take up the clearance between a piston and its bore, called *knurling*. Depressions are put in the thrust faces of a piston, creating raised projections which increase a piston's diameter and reduce its bore clearance. The problem is, knurling is *temporary*. The projections wear off quickly, putting the wear right back where it was before all your work. Don't do it! Replace your piston/s if their clearance exceeds 0.008 in.

Glaze-Breaking—If your cylinder bores are within acceptable taper limits, it is good practice to *break the glaze* in the cylinder bores. The glaze-breaking operation doesn't remove any appreciable material from the bore, it merely restores a hone-type finish for positive ring break-in. If you are using plain or chrome rings, this operation is necessary. It is optional with moly rings—but desirable because of the ring-break-in aspect. Again, the object of glaze-breaking is not to remove material from a bore, therefore a precision-type hone should not be used because it will try to remove what bore taper is there, thus increasing piston-to-bore clearance. A spring-loaded or ball-type hone that *follows* the existing bore without changing

Glaze-breaking, or restoring cross-hatch to bores that don't require reboring is particularly important if chrome rings are to be used.

Running a large flat file over the block's deck surface will remove any imperfections that may cause this style boring bar to be mislocated. The boring bar removes only enough material to clean up the bore and to accept the next largest oversize pistons. 0.002"–0.003" stock is left for honing.

its shape should be used to break the glaze on the bores.

DECKING YOUR BLOCK

Before sending your block off to be bored or honed, first check for any irregularities on the deck surface if either one of the head gaskets showed signs of leakage. If you have additional machining operations done on the block, you'll want to have them done *all at one time.* Another, more important reason is many engine machine shops use a boring bar that locates off the block's deck surface. If the deck surface is off, the new bore will also be off relative to the crankshaft bearing-bore center.

If your block is *notched* between two cylinders, it may have to be welded if more than the following amount has to be removed to clean the deck surface up: 0.015 in. from 221s through pre-1977 302s; none from '77 and later 302s; 0.020 in. from Boss 302s; 0.034 in. from pre-1977 351Ws and 0.014 in. from '77 and later 351Ws. Removing more than the amounts indicated will cause deck-height clearance problems. Another disadvantage with milling a block's deck surface, particularly these days, is the engine's compression ratio will be increased slightly, raising the engine's octane requirements. High-octane fuels are already hard to find—not to mention the added cost.

Because the block is cast-iron, a special welding process is required to repair it. The block must be preheated prior to welding, and a nickle-alloy rod used to do the welding. After the notch is welded, the raw weld must be machined flat, or made flush with the rest of the block's deck surface. Although it's not your problem, this is not as easy as it sounds because nickle-alloy weld is very hard and nearly impossible to machine, consequently it must be ground. *Your* problem is the cost of having it done. Compare this cost to that of a replacement block.

For less severe surface irregularities, check your block if there was any sign of gasket leakage. To do this, you'll need an accurate straight-edge at least 6-inches long and some feeler gauges. Check for any gaps between the straight-edge and the block surface for any warpage. If your straight edge extends the full length of the block, the maximum allowable gap is 0.007 in. However, if your straight-edge is 6-inches long, the maximum is 0.003 in. These two figures equate to basically the same amount of warpage. If you do find warpage, check the figures a couple of paragraphs back and do not have more than the maximum amount machined off the head-gasket surfaces.

A final thought about surfacing your cylinder block. If one deck surface requires surfacing, the other one will have to be done whether it needs it or not. This maintains equivalent deck heights from side-to-side.

Other Checks—Other than checking your block for cracks, amount of bore taper and deck flatness, there's not much more to look for. In the normal course of things, nothing else wears out or can go wrong with the cylinder block. For example, the valve-lifter and distributor bores are lightly loaded and so drenched in oil that they just don't wear. The only problem would occur if they were to be damaged from something such as a broken connecting rod which somehow got into the cam and lifters. This would require junking the block. To be sure lifter bores are OK, check them with a *new* lifter. Do this without oiling the lifter or the bore. Insert the lifter in each bore and try wiggling it—if there is noticeable movement, the bore is worn and the only way of fixing it is to replace the block. Sleeving the lifter bore is possible, but difficult to have done.

CYLINDER-BORE FINISHING

Plain Cast-Iron, Chrome, or Moly Rings?— When you drop your block off to have it bored or honed, you should know the type of rings you intend to use so you

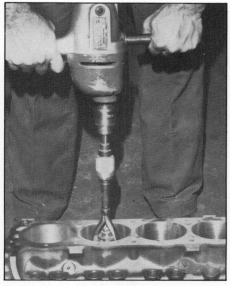 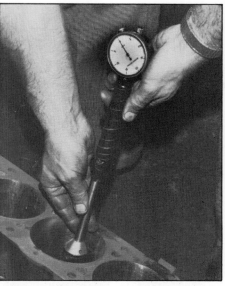 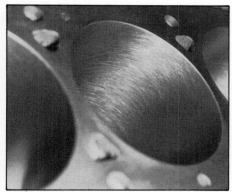

Cross-hatch pattern for ring seating and oil retention.

Honing the block after boring with a hand-held hone. Bore diameter is checked periodically with a dial-bore gauge. Main-bearing caps must be in place and torqued to spec to produce accurately honed bores.

can tell the machinist. Your engine's final bore finish will be different depending on whether you use plain, chrome or moly rings. Chrome rings are especially suited to engines that will inhale a lot of dust and dirt such as a truck used in a rock quarry. Chrome is very tough and lasts longer than the moly type under these conditions. However, unless you have an extraordinary situation, use moly rings. Their life will be much better than the chrome type and they require virtually zero break-in mileage. As for plain cast-iron rings, they break in quicker than *chrome-faced cast-iron rings,* but they also wear out quicker. I recommend you don't consider them.

What's the Difference?—The advantage of a moly ring over a plain or chrome one is the moly-type ring carries more of its own oil. It has surface voids, or little depressions, which contain oil much in the same manner as the cross-hatching of a honed cylinder. Plain and chrome rings have virtually no voids and must depend almost totally on cylinder wall cross-hatch to supply lubricating oil. When the piston travels down on its power stroke, the cylinder wall is exposed to a burning fuel charge. Consequently, the oil on the wall is partially burned away, meaning the piston rings will not receive their full amount of lubrication during the return trip back up the cylinder on the exhaust stroke. The moly-type ring carries its own lubrication which is not directly exposed to the combustion process.

The reason for different bore finishes to match the ring type should now be obvious. A chrome ring depends on a coarsely finished cylinder wall to hold lubricating oil. If plain or chrome rings are used, the bore should be finish-honed with a 280-grit stone. A 400-grit stone is used for moly rings. A 30° cross-hatch hone pattern is used for all ring types.

Install the Main-Bearing Caps—Prior to delivering your block for boring and/or honing, install the main-bearing caps. Torque them to specification: 95—105 ft. lbs. for the 351 and 60—70 ft. lbs. for the 221 and through the 302 engines. Outer bolts for the Boss 302 are torqued 35—40 ft. lbs. This is necessary because the load imposed on the cylinder by the main-bearing bolts distorts their bores slightly. Therefore, the bores have a different shape when the block is bare than when the main caps are installed. The object is to simulate the stress and deflections in the block so the machining processes will provide bores as close to perfect as possible—after the engine is assembled. The cylinder head is another source of cylinder stress and deflection, but how can you hone the engine with a head installed? Fear not, the dilemma has been solved. Some engine machine shops use a *torque plate* when boring and/or honing. The 2-inch-thick steel torque plate has four large-diameter holes centered on the engine bores to allow boring or honing. The plate is torqued to the block during these operations. If the shop you choose does not use a torque plate, don't be concerned. This is not absolutely essential unless the engine is intended for all-out racing. Also, using a torque plate increases the machining costs because it takes time to use it. If you are on a tight budget, make some cost comparisons. Regardless of whether or not a torque plate is used, *do* install the main bearing caps.

Chamfer the Bores—When you get your block back from the machine shop, inspect the bore tops. The machinist should have filed or ground a small chamfer or bevel at the top of the bores after honing them. The reason is to assist the piston installation. It will provide a lead-in for the piston and the rings, and it will also eliminate a sharp edge that will get hot in the combustion chamber. A 45°, 1/32-in.-wide chamfer is sufficient. A fine-toothed *half-round* or *rat-tail* file works well for this job. Just hold the file at a 45° angle to the bore as you work your way around the top of each cylinder. Don't hit the opposite side of the cylinder wall with

After boring and honing a sharp edge is left at the top of each bore. Chamfer the bores with a file to provide a lead-in for installing the rings and pistons.

The most important job after boring and honing is to clean your block. A steam cleaner is being used here after hot tanking. A scrub brush in the bores with some laundry detergent, a bucket and a garden hose works equally well.

Dry your block off with some paper towels immediately after rinsing. Protect the machined surfaces against rust by coating them with oil.

the end of the file. You don't want to gouge your freshly machined bore.

Clean It Again—Even though your block was hot-tanked, spray-jetted or whatever method you used for cleaning it, your block must be cleaned again. If you are trying to save your cam bearings, inform your machinist, otherwise he'll automatically clean the block after he's machined it. If you intend to save the bearings and he does his cleaning with a hot tank, goodbye bearings. This second cleaning removes machining residue, mainly the dust and grit left from the honing stones. If it isn't removed before the engine is assembled, your engine will eat up a set of rings so fast it'll make your wallet ache. Grit will also be circulated through your engine's oiling system and end up embedded in the crankshaft, connecting-rod and camshaft bearings, turning them into little grinding stones. Consequently, even though your engine was hot-tanked by the machinist, enlist the use of a stiff-bristle brush, laundry detergent and some warm water to satisfy yourself that the engine is clean. This combined with generous amounts of elbow grease will give you a super-clean block. Wiping the bores with a white paper towel is a good check. It should stay clean and white.

After you've finished the scrubbing and rinsing, *don't let the block air dry*. The newly machined surfaces will rust very quickly. If you have compressed air, use it to blow-dry the block. It's best because you don't have to touch the block and the compressed air will blow water out of every little corner. The next best thing to use is paper towels. Don't use cloth rags or towels because they spread lint throughout the engine—lint that won't be dissolved like that from a paper towel. *Immediately* after you have the block dried off, get out your can of spray oil and coat all the machined surfaces. CRC® and WD-40® displace water rather than trapping it. This will protect the block against rust, but will also turn it into a big dust collector. Consequently, you'll have to keep it covered up. Because you're finished with the block until it comes time to assemble the engine, store it by standing it on its rear face and cover it up with a plastic trash bag. This'll keep the block clean and dry until it's time for final engine assembly.

THE CRANKSHAFT

The crankshaft is the main moving part in any engine. Most of the other parts turn the crank, support it and lubricate it. Because of its importance, a crankshaft is a very tough, high-quality component. As a result it's rare that a crankshaft cannot be reused. They very rarely ever wear to the point of having to be replaced. What usually happens is the journals are damaged from a lack of lubrication, but mostly from *very* dirty oil. Usually this damage can be repaired. Even dirty oil usually doesn't damage a crankshaft's journals, but if the dirt particles get big and numerous, it will. The one thing that can render a crankshaft useless is mechanical damage caused when another component in an engine breaks—such as a rod bolt. If an engine isn't shut off immediately when this happens, or if it occurs at high RPM, the crank and many of the other expensive components will have to be replaced also.

Main and Connecting-Rod Journals—The most important crankshaft inspection job is to make sure the bearing journals are round, their diameters don't vary over their length and their surfaces are free of imperfections such as cracks. A cracked crankshaft *must* be replaced. Except for the Boss 302, all small-block Ford cranks are made from cast iron, and if you've found a crack, consider yourself lucky. A crack in a cast-iron crankshaft is followed very quickly by a break. So don't even think about trying to save your crankshaft if it is cracked. The best way of checking for cracks is by Magnaflux®, the next is by dye testing and the last is by visual inspection. Check with your local parts store or engine shop about having your crank crack-tested, preferably by Magnaflux®. The cost is minimal because a broken crank can destroy an engine.

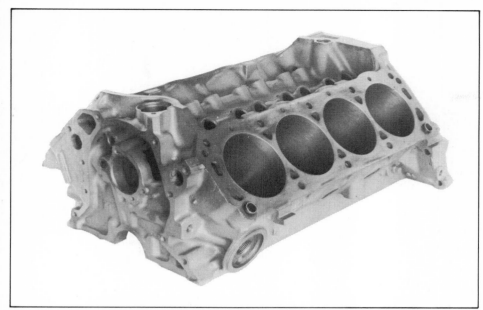

Your block should look something like this now. Cover it up with something like a plastic trash bag to keep it clean and dirt free until assembly time.

If a journal is oval or egg-shaped it is said to be *out-of-round*. This condition is more prevalent with rod journals than with main-bearing journals. If a bearing journal's diameter varies over its length, it is *tapered*. This is like the taper of a cylinder bore except it's an outside diameter. Finally, the journals should be free of any burrs, nicks or scoring that could damage bearings. Pay special attention to the edge of each oil hole in the rod and main journals. Round these edges if you find any sharp ones. Sharp edges here can cut the bearing material out of the rod and main inserts.

Beginning with the 221, there have been many subtle changes to the small-block Ford, however the crankshaft's excellent initial design has remained virtually unchanged. For example, the nominal (average) diameter for the main- and rod-bearing journals have remained at 2.2486 in. and 2.1232 in. respectively with exception of the 351. Its journals were increased to 2.9998 in. for the mains and 2.3107 in. for the rod journals for additional bearing capacity for the larger engine. The nominal size for all the journals can vary ±0.0004 in. and be within specification.

Surface Finish—Look at surface finish first. If any of the bearing journals are rough, they'll need to be refinished or

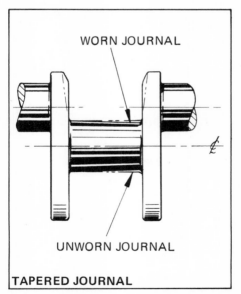

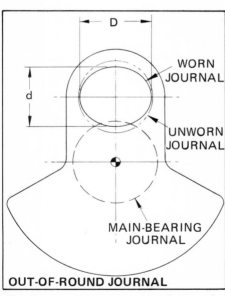

Bearing journals wear two ways, tapered or varying diameter over their length and out-of-round or oval in shape. Out-of-round is more prevalent with connecting-rod-bearing journals.

Deep grooving on this connecting rod journal from foreign material in the oil. This crankshaft should be reground or replaced.

Measuring connecting-rod and main-bearing journals for wear. Record your measurements so you'll have something to refer back to if you need to recheck.

reground regardless of taper or out-of-roundness. About the only way of determining if your crankshaft journals are *smooth* is to run your fingertips over them. If a journal looks rough and you can feel it, the bearing will also *feel* it. A regrinding job will then be in order. If it is required, the normal 0.010-in. regrind job will hopefully take care of the journal surfaces and any normal wear. I say *surfaces* because if one journal has to be ground, all rod or main journals should be ground the same amount. You don't want *one* odd-size bearing—bearings are sold in sets. The additional cost will be slight because the major expense is incurred in the original setup. Therefore, if 0.010 in. is removed from the standard-size bearing journals, 0.010-in *undersize* bearings will be required. Undersize refers to the crankshaft journal. The amount of undersize is the difference between the journal diameter after being reground and the nominal or average specified standard-size journal diameter. For example, if the nominal-specified diameter for the main journal of a crank is 2.2486 in. and it is reground to a diameter of 2.2386 in. it is 0.010-in. undersize.

While you're checking bearing journals, take a good look at the thrust faces on the number-three main-bearing journal. This is particularly true if your engine is backed by a standard transmission. When a clutch is disengaged, the release bearing pushes forward on the crankshaft. This load is supported by the crankshaft's rear thrust surface, so if the pressure plate had an excessively high load or your clutch was wrongly adjusted, the thrust surface may be damaged. An indication as to whether you'll have thrust surface trouble is the condition of your pressure-plate release levers. If they show signs of excessive wear from the release bearing then be particularly attentive. Measure between the thrust faces. You'll need a snap gauge and a micrometer or an old or new bearing insert, feeler gauges and some mikes. Minimum width is 1.137 in.; maximum is 1.145 in. Check the thrust-face widths in several places. If the 1.145-in. figure is exceeded, you'll have to replace your crank or have it reconditioned.

Out-of-Round—To check for out-of-round or tapered journals you have to have a 2- to 3-in. outside micrometer and an understanding of what you're looking for. If a journal is round it will be described by a diameter which you can read directly from your micrometer. If it is out-of-round you will be looking for the *major* and *minor* dimension of an ellipse which can also be read directly. An out-of-round journal will be a close approximation of an ellipse. The minor dimension can be found by measuring around the journal in several locations. The major and minor dimensions will occur about 90° from each other. The difference between these two figures will be how much a journal is out of round, or

Major Dimension — Minor Dimension = Out-of-Round

Start with the connecting-rod journals when checking. They are most likely to suffer from this out-of-round condition because they are highly loaded at their top and bottom dead centers. With this in mind, check each journal in at least a couple of locations fore and aft on the journal. The maximum allowable out-of-round is 0.0004 in. for both crankshaft and rod journals. Remember, if you find *one* journal out of spec, your crankshaft should be reground.

Taper—Crankshaft journal taper causes uneven bearing wear more than an out-of-round journal. This is due to uneven bearing load over the length of the bearing. As a journal tapers it redistributes its load from less on the worn portion or smaller diameter, to more on the less-worn part or larger diameter of the journal. This results in rapid bearing wear when a crank with excessively tapered journals is reinstalled with new bearings.

Taper is specified in so many tenths of thousandths-of-an-inch per inch. What this means is your micrometer readings will vary by this amount when they are taken one inch from each other in the same plane.

For connecting-rod journals, maximum allowable taper is 0.0004 in./in. and for main-bearing journals it is limited to 0.0003 in./in. Again, if one journal is out of spec, all the other journals should be reground at the same time the faulty journal is being corrected.

If your crank needs regrinding, trade it in on a reconditioned one. You'll get someone else's crankshaft that has been reground. The name for this is a *crank kit*. It consists of a freshly reconditioned crankshaft *and* new main and connecting-rod bearings. I recommend taking this approach rather than having your crankshaft reground, *and* buying new bearings. You'll end up saving about $25. Just make sure the bearings accompany the crankshaft.

After Regrinding—Check your crankshaft carefully if you've had it reground or traded yours in on a reconditioned one. The two main things to look for are sharp edges around the oil holes and improper shoulder radii at the ends of the journals. Oil holes should all be chamfered, or have the edges removed to prevent the bearings from being damaged. As for the radius, there should be one all the way around each crank journal. If there's a corner rather than a radius, it will weaken the crankshaft because the corner is a weak point, or a good place for a fatigue crack to start. A radius that's too large is great for the crankshaft, but bad news for the bearings. The bearing will *edge ride,* or ride up the radius rather than seat in the journal. Use the accompanying sketch to make a *checking template* for checking journal radius. Make it out of cardboard—the type that comes inside your shirts from the cleaners—or some thin sheet metal if you want something more durable. The maximum *specified* main-journal and connecting-rod-journal radius is 0.100 in. If any crank radius isn't right, return the crank to be corrected, otherwise it could spell real trouble.

Smooth and Clean Journals and Seal Surfaces?—A newly reground crankshaft should have its journals checked for roughness for no other reason than they may be too rough—just like you checked your old crankshaft, but not the same way. Rather than using your finger, rub the edge of a copper penny lengthwise on the journal. If it leaves a line of copper, the journal is too rough. If this is the case, consider smoothing the journals rather than going through the hassle of getting it redone. More often than not, you'll end up with a better job by doing it yourself without going through the frustration which usually accompanies getting a job redone. Also, if your original crankshaft checks out OK in all departments, you should give it the same treatment just to put a *tooth* on the highly polished journals and oil-seal surface in addition to removing any varnish buildup.

Regardless of whether the journals are newly reground and need smoothing or are original, polish them with 400-grit emery cloth. A 1-inch-wide strip a couple of feet long should be sufficient to do all the journals and the rear-main seal surfaces. Wrap the cloth around the journal as far as possible and work it back and forth lightly as you gradually move around the journal. Keep track of where you start on a journal so you can give it an even finish all around. If you stay in one spot too long or use too much pressure, you'll remove material unevenly. The object isn't to remove material, at least as little as necessary, but to give the journal and rear-main-seal surfaces a clean smooth surface. Be careful with the oil-seal surface. Don't do too good a job. Polish it just enough to remove any varnish or to smooth out any nicks or burrs. Some oil is required between the seal and the crankshaft to lubricate both and a highly polished surface will seal too completely with the result being eventual seal failure.

Crankshaft Runout—Crankshaft *runout* describes how much a crankshaft is *bent*. It is found by rotating the crank between two centers and reading runout with a dial indicator set 90° to each main-bearing journal. As the crankshaft is rotated, the indicator reading will change if the crankshaft is bent.

I haven't mentioned runout until now because it's not likely you'll encounter this problem. If you do it may be a false alarm because a crank just laying around on the garage floor for some time can change. Not enough that you can see, but enough to show with a dial indicator.

Dial-Indicator Checking Method—All you'll need is an indicator with a tip extension so the crank throw won't interfere with the dial-indicator body as the crank is rotated. Just set the crank into the top halves of your old front and rear bearings with no oil seal. Leave out any of the intermediate top bearing halves. Oil the bearings well. This leaves the crankshaft free to wobble in the center as you turn it. Mount your dial-indicator base to the block with the indicator at 90° to the main-bearing journal being checked. Offset the indicator tip on the journal to miss the oil hole. Rotate the crankshaft until you find the lowest reading and zero the indicator dial. You can turn the crankshaft

With only the front and rear main bearings and caps in place and torqued to spec, total indicated reading obtained with a dial indicator set up as shown should not exceed 0.004 in.

and read runout directly. Turn the crank a few times to make sure you have a good reading. Maximum allowable runout is 0.004 in. If your crank exceeds this, don't panic. As I said, a crank can change from just sitting around, so if yours is beyond the limit, turn the side that yielded maximum runout up, or down as the crankshaft would look with the engine in its normal position. Install the center main cap with only the bearing in the cap. This will *pull* the crank into position. Leave it this way for a day or two and re-check the runout. Your crankshaft will probably *creep,* or bend to put it within the 0.004-in. range. However, if it's too far off and can't be corrected using this method, regrind it or trade it for a crank kit.

Cleaning and Inspecting the Crankshaft—When cleaning a crankshaft the most important thing is to concentrate on the oil holes. Get them really clean. Even if you sent your crank along with the block for cleaning, some wire tied to a strip of cloth pulled through the oil holes will remove what's left if the hot tank didn't get it all. Soak the rag in carburetor cleaner or lacquer thinner and run it through each hole several times. If you want to be sure

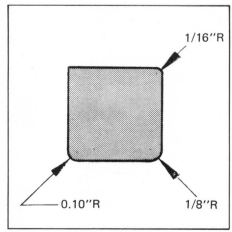

Use this pattern for making a template to check bearing-journal radius. A journal radius should be no larger than 0.100 in. otherwise it will edge-ride the bearing. The 1/16-in. (0.0625-in.) and 1/8-in. (0.125-in.) radii are for reference.

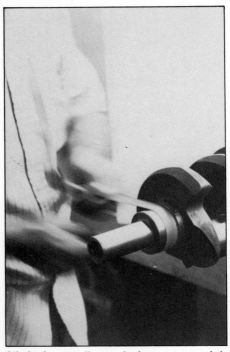 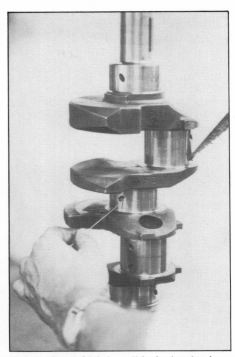 

All that's normally required to put a crankshaft back into shape: Lightly polish the bearing journals with some fine-grit abrasive cloth, clean up with special attention to the oil holes and rust-proof the machined surfaces.

the oil holes are really clean, a copper-bristle gun-bore brush is great for the job. You may have used one of these for cleaning the block oil holes. By all means use it if you have one. The oil holes are the only thing needing cleaning at this point, however if the hot tank, or whatever you used, didn't get all the residue off the throws and counterweights, a stiff brush and solvent can be used to get what remains.

Installation Checking Method—To do a real-world check, install the crankshaft in the block using oiled *new* bearings, but without a rear-main seal. Torque the caps to specification. If the crank can be rotated freely by hand, consider it OK. Any loads induced by what runout there is will be minor compared to the inertial and power-producing loads normally applied to the journals and bearings when the engine is running. I suggest this way of checking because crankshaft runout isn't usually a problem with the normal "tired" engine that just needs rebuilding. If you decide to use this check, follow the procedure for crank installation from the engine assembly chapter.

Pilot Bearing—With a standard transmission, a *pilot bearing* or bushing will be installed in the center of the flywheel flange. It supports the front of the transmission input shaft which pilots into it, thus the name pilot bearing. Visually check its bore for damage. If you have an old transmission input shaft, insert it into the bearing and check for lateral play by trying to move the shaft side-to-side. If there is noticeable movement, the bearing should be replaced. For a more accurate check, measure its bore with a snap-gauge and micrometer. The bearing's standard bore diameter is 0.625 in. and its maximum allowable diameter is 0.628 in.

To be on the safe side I recommend that you replace the pilot bearing and don't bother to check it. The cost is minimal and it's a lot easier to do now rather than after your engine is back in place and in operation. You can remove a pilot bearing by using any one of four different methods—there are more, but these are the most common ones. Two of the methods require special tools—a *puller* or a *slide hammer*. The slide hammer shaft screws into or hooks onto the back of the pilot bearing and has a heavy sliding handle. Sliding the handle backwards bottoms it against the end of the shaft. It will eventually knock the bearing free. The puller screws into the pilot bearing and uses a steady force to pull the bearing rather than an impact force. Tightening a nut on the puller gradually pulls the bearing out of its bore. Two homemade methods employ the same basics. The first involves filling the pilot-bearing bore with grease. To remove the bearing, insert either an old input shaft, a short 5/8-in. diameter bar or a bolt (with its threads cut off) in the bore and hit it with a hammer. Hydraulic pressure will force the bearing loose eventually. The final method uses a coarse-threaded 11/16-in.-diameter bolt ground with a long taper on its end—the first three or four threads. The bolt can be threaded into the bearing until it bottoms on the crankshaft. Keep turning the bolt until it pushes the bearing out. Regardless of the method you choose to use, make a note of how deep the bearing was located in the end of the crank and replace it in the same position.

Prior to installing your new bearing, soak it in oil. The bearing is made from Oilite, a special bronze which is self-lubricating, but soaking it in oil gives it additional lubrication. Also grease the bearing bore in the end of the crankshaft to assist in installing the bearing. Line the bearing up with the crank bore and tap the bearing into place. Use a brass mallet and a thick-walled tube or pipe which fits the outer edge of the bearing to prevent distorting the bearing's bore as it is driven into place. When starting to drive the bearing in, go easy at first being careful not to let the bearing get cocked. Once it is started straight, it should go the rest of the way. If it does cock, stop and remove it, then start over. Drive the bearing in until it lines up with the bottom of the counterbore in the center of your crankshaft. Check the bearing by putting the transmission input shaft into the bearing bore. The shaft should enter and turn easily without binding.

CAMSHAFT AND LIFTERS

After leaving a very tough customer like the crankshaft, we now come to one not so tough—the camshaft and its lifters. As opposed to a crankshaft, once a cam starts wearing it's usually not long before the

A puller being used to remove a pilot bearing from the end of a crankshaft. The flywheel is on this crankshaft because the bearing is being replaced as part of a clutch replacement and not an engine rebuild.

A new bronze pilot bearing. Be careful when installing yours so as not to distort its bore. Drive it in until it bottoms squarely in the crankshaft.

cam lobes and lifters are *gone*. Consequently, you must give your old camshaft and lifters a very good look before deciding to reuse them. Two hard and fast rules apply to camshaft and lifter combinations:

1. When reusing a camshaft and its lifters, **never install the lifters out of order.** The lifters must be installed on the same lobes and in the same lifter bores.
2. **Never install used lifters with a new camshaft.**

Almost without exception, the rules related to proper cam and lifter combinations and installation methods have to do with avoiding excessively high contact pressure between cam lobes and lifters. The reason is, loads exerted between a cam and its lifters during normal or acceptable operating conditions are extremely high. Consequently, if there's any wrong move made between now and the first 30 minutes of engine run-in, the chances of a camshaft and its lifters being ruined are very high. Therefore, it's important to adhere to proper procedures when choosing and installing your camshaft.

Camshaft Lobe and Lifter Design—To have a clear understanding of why certain things have to be done when dealing with your engine's camshaft and lifters, a quick education in cam lobe design is in order. First is the *profile* of a lobe, or what one would look like when viewing it looking at the end of the cam. The cam profile governs how much the valves open and when they open and close. When a valve is closed, the lifter is on the *base circle* of the cam lobe described by a radius. When the lifter is on the highest point of the cam lobe, it's on the *toe* of the lobe and the valve is fully opened. The accompanying sketch shows the difference between the *base-circle diameter* and the distance from the toe directly across the camshaft centerline to the base circle is the cam's lift at the lifter—*not at the valve*.

During the camshaft manufacturing process, lobe surfaces are not ground parallel to the camshaft centerline, but are ground at a *rake angle*—1° in most instances. Also, the lifter *foot* that contacts the cam, is not ground flat but has a *spherical radius,* or is ground *convex*. Ford used a 30-in. radius. At first this sounds odd because one would think this would reduce contact area between the lobe and lifter. This machining ensures reasonable contact pressure and good lubrication within practical manufacturing tolerances. The lifter-foot radius and lobe angle guarantee good cam-to-lifter contact area for consistent camshaft and lifter life.

The way camshafts and lifters are machined serves another important function. Cam-lobe contact is made *off-center* rather than directly on the lifter's center. Sliding friction between a cam lobe and its lifter rotates the lifter, minimizing wear of the two components.

CAM AND LIFTER INSPECTION

During their operating life, a cam and its lifters gradually wear. How much depends on how an engine is operated and, more importantly, how it is maintained. You must determine whether or not this wear has progressed to the point that yours shouldn't be used. I'll say it right now. Using an old camshaft and lifters in a newly rebuilt engine is a risky deal. It's not uncommon for a new cam to fail during its first 100 miles of operating in a newly rebuilt engine, let alone a used one. Finally, if you've lost track of the order of your lifters, toss the lifters away right now and trade your cam in on a reground or new cam and new lifters. The odds of getting lifters back in the right order onto 16 cam lobes are absolutely astronomical: 20,922,789,890 to 1.

Check the Camshaft First—When checking your cam and lifters, you check the cam first because if it's bad you'll have to replace the cam *and the lifters* regardless of the condition of the lifters. Remember the second rule, **never install used lifters with a new cam.**

The first thing you do is get out your trusty micrometers—vernier calipers are OK—and check your cam's *lobe lift*. Maybe you already did this during the diagnosis process prior to removing and tearing down your engine. You'll have used a dial indicator at the end of a pushrod to check the lobe lift that the rocker arm sees, or the *actual* lobe lift. If you've already used this method and have made a determination about your cam you can forget measuring the lobes directly because lift measured at the lifter or pushrod is accurate.

What you are looking for when measuring camshaft lobes with the cam out of the engine is not so much their lift as a *comparison* between their lifts. As I said previously, when camshaft lobes begin to wear, they don't do it at the same rate. Consequently, every lobe but one may be in good shape. To measure a cam lobe you have to measure each lobe in two places. First, measure the lobe's base-circle diameter 90° to the line that goes through its toe. This will give you the lobe's *minor dimension*. Next, measure the *major dimension* on the line through the toe and cam centerline. Make both of these measurements where there is maximum visible wear and the same distance from the same edge of the lobe. With these two dimensions:

Major Diameter — Minor Diameter
= Approximate Lobe Lift

CAUTION: This method of determining camshaft lobe lift becomes increasingly

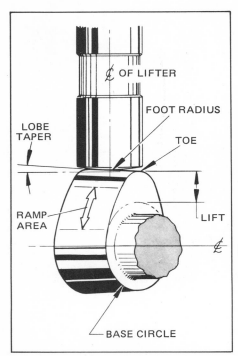

A visual inspection of the camshaft-lobe wear pattern and the lifter foot will tell you if your cam should be replaced. If the wear pattern is the width of the cam-lobe toe and the lifter foot is worn flat or concave, your camshaft *and* lifters need replacing.

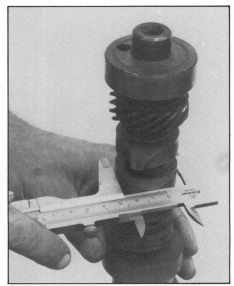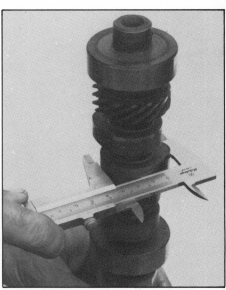

The difference between these two measurements will give you camshaft *lobe* lift. It should measure approximately 0.23"–0.30" depending on which engine you have. This method is not accurate for determining high-performance cam lift. The ramp area of a long-duration cam extends farther around the base circle resulting in a larger-than-actual base-circle diameter reading, or a less-than-actual lobe lift. See chart on page 9 for lobe lifts.

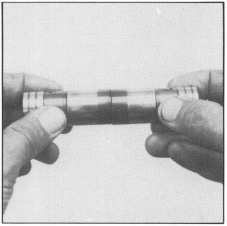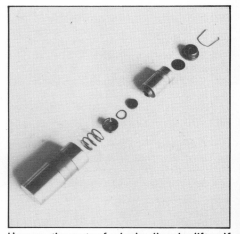

Putting the foot-end of two of your worn lifters together will tell you if any spherical radius exists. If you can't rock them relative to one another, they can't be reused and your cam and lifters must be replaced.

Here are the parts of a hydraulic valve lifter. If you disassemble yours for cleaning, make certain you don't lose any parts or interchange any plungers between lifter bodies.

inaccurate with high-lift, high-performance camshafts.

After arriving at a figure for each lobe compare all the intake lobes, then compare all the exhaust-lobe figures. If the figures aren't within 0.005 in. you know some lobes are worn excessively and the cam should be replaced. As for which of the lobes are for intake and which are for exhaust valves, from either end of the camshaft they are:
I E E I I E E I I E E I I E E I.

Another thing to notice is the wear pattern on each lobe, particularly at the toe. If the wear has extended from one side of the toe to the other, the cam has seen its better days and should be replaced, even though its lift checks OK. This goes for pitting too. *Any* signs of pitting on the *lift areas* of a lobe indicates metal loss which will probably show up on its mating lifter. A "full-width" wear pattern or pitting of a cam's lobes indicates the cam and lifters should be replaced.

Cam Bearing Journals Are "Bullet-Proof"—As for bearing journals, they never seem to wear out—at least I've never seen any worn out. However, if you feel compelled to check yours while your micrometer is handy, do it. Checking never hurt anything and the key to building a good engine is check, check and recheck. Just like the crankshaft, there are five camshaft bearing journals, but unlike the crank, they are all different sizes. Numbering the journals from front-to-back, the limits for all small-block bearing-journal diameters are:

Bearing Number	Journal Diameter (inches)
1	2.0805–2.0815
2	2.0655–2.0665
3	2.0505–2.0515
4	2.0355–2.0365
5	2.0205–2.0215

Cam journals also have a maximum out-of-round and runout spec like the crank, a 0.0005-in. limit for out-of-round and a 0.005-in limit for runout. Again, it's unlikely these limits will be exceeded unless your engine has been "gone through" previously and a many-time-recycled cam installed. If this is the case, you'll be better off replacing the cam anyway. When checking cam-journal runout, use the same method used for checking crankshaft journals—find maximum and minimum dimensions and subtract them. As for runout, you can't check it unless you have centers to mount and rotate the cam on or a surface plate and some angle blocks in addition to a dial indicator. However, don't worry about it. Cam runout is the least of your worries because it seldom happens. The real check is if it turns easily when installed in the block. If it does, consider it OK.

Check Your Lifters If the Cam Is OK—I may sound like a broken record, but be careful not to get the lifters out of order. If you do, you must pay the price and replace the lifters, even if the cam checks

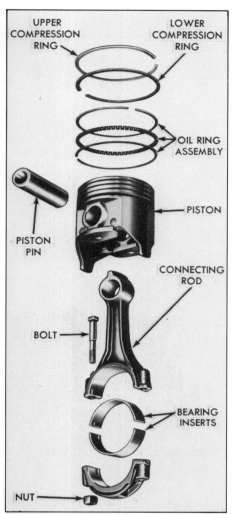

Complete piston and connecting-rod assembly including bearings and piston rings. Wrist pin is retained in the assembly by an interference fit between the pin and the connecting-rod pin bore. Drawing courtesy Ford.

lift checks out OK. IF this is the case, chances are the wear pattern shows all the way across the toe of the lobe this lifter was mated with.

Clean the Lifters—Concern yourself with cleaning your lifters only if your cam is reusable and the lifters are OK. Clean the varnish buildup from around the foot of the lifters and any sludge or varnish which has accumulated inside the lifters if they are the hydraulic type. The outside will clean by soaking the lifters in carburetor cleaner available at your local automotive parts store. Soaking doesn't get the inside at all. You can take two approaches here.

The first and easiest method of cleaning a hydraulic lifter is to use an oil squirt can to force lacquer thinner through the hole on the side of the lifter body as you work the plunger up and down. The most positive method of doing it is to take the lifter apart, but it's also the most difficult method. There are eight parts to contend with, starting with the retaining clip inside the groove at the top of the lifter body. Be careful during disassembly so you don't damage the lifter components or

lose any of the little parts. Disassemble and clean one lifter at a time so there's no chance of mixing up the parts because they are matched sets. You'll need small needle-nose pliers and a lot of patience, so much that you may decide to revert to the lacquer thinner and squirt-can approach. If you do decide to continue with this method you may discover a need to use the squirt can to help loosen up the plunger so it will come out of the lifter body. Do this after you've removed the retaining clip, pushrod cup and metering valve. By working the plunger up and down with a few injections of thinner into the body, the plunger should gradually work its way out. Besides the plunger, you'll have more parts to contend with as the plunger comes out. They are easy to lose, so be careful.

After cleaning each of the lifter's components separately, reassemble that lifter *before* starting on another one. This prevents mixing components and it's easier to keep the lifters in order. To assemble a lifter, stand the plunger on your work bench upside down and assemble the parts that go below it in the

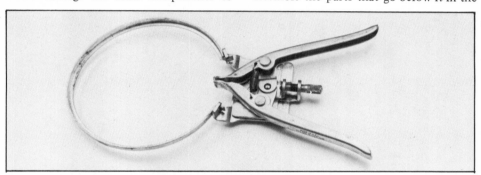

A ring expander is handy for removing or installing piston rings.

OK. Of course, you won't have to bother checking the lifters if your cam is not reusable. If you can reuse the cam, proceed.

The thing you're looking for when checking lifters is the spherical radius at the foot. If the radius is gone, the lifter is junk. Because this radius is so large—about 30 inches—it's difficult to check it with a straight edge, even if the lifter is brand new. So double the radius by butting two lifters together end-to-end. They will rock back-and-forth noticeably if any radius is left. If they won't rock, the lifter is either worn flat or concave and the lifter is junk. I hate to keep saying this, but because of the touchy nature of a camshaft you must be aware of the consequences. If the lifter is worn concave, not only the lifter should be replaced, but the cam also—even though

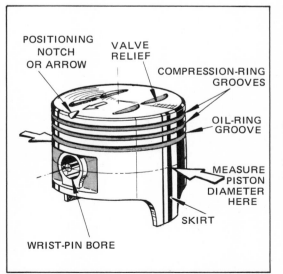

In addition to the piston nomenclature, note the notch or arrow in the top of the piston. It is important for assembling a piston to its connecting rod, then for installing the complete assembly in its bore.

When measuring a piston, do it level with the wrist-pin and 90° to its axis. Width measured across the bottom of the skirts should be slightly larger—about 0.0005 in. One thing is wrong with the picture. The piston and connecting rod has been disassembled. This may not be necessary, so don't do it.

Scuff marks on a piston's skirt indicate the engine has been severely overheated, resulting in the skirts being overstressed and possibly collapsed.

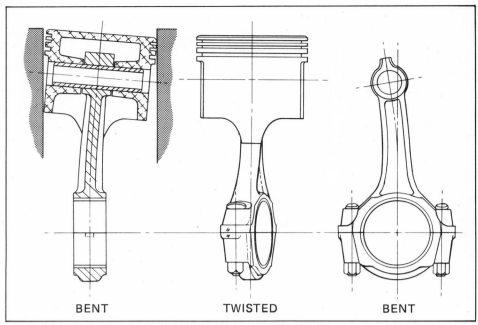

The first two rods, bent and twisted as shown in these exaggerated drawings, cause uneven piston skirt wear as the pistons are tilted in their bores. The first rod, with its wrist-pin axis bent out of parallel with its bearing-bore axis, tilts the piston mostly at TDC and BDC. The twisted rod does its tilting progressively toward the mid-point of its stroke. A connecting rod bent like the third one shown doesn't affect piston wear or engine performance. Consequently it can be reinstalled unless the bend is obvious to the eye or it's for an all-out performance engine.

lifter body on the plunger. Depending on the type lifter you have, there will be in order a check valve, a small check-valve spring, and/or a check-valve retainer and a plunger spring. With these stacked up on the plunger, oil the ID of the lifter body and slide it down over the plunger, check valve and spring assembly. You can now turn this assembly over and install the metering valve and disc or just a metering valve, again depending on the type lifter you have, the pushrod cup and retain the complete assembly with the retaining clip. One lifter cleaned and assembled. Do this 15 more times and you're finished cleaning your lifters.

PISTONS AND CONNECTING RODS

Replace the Pistons?—If your block has to be rebored for one reason or another, you'll have to replace your pistons—the holes are too big now. However, if you determine from the following that most of your pistons should be replaced, then the block should've been bored in the first place. The cost of pistons is the major expense, so you may as well give your engine a new start with larger pistons and new straight bores. Also, its durability will be as good or better than new, depending on the care you take during the course of the rebuild.

CHECKING THE PISTONS

If your engine doesn't have to be rebored then your next step is to check the pistons to establish whether or not they are reusable *before disassembling them from the rods*. Remove the old rings from the pistons and toss them away. Be careful when doing this so you don't scratch the pistons. A *ring expander* will help. Don't remove the rings over the skirt, remove the top ring first, then the second followed by the oil ring. If you do this by hand, make sure the ends of the rings don't gouge the piston. For convenience, support the piston-and-rod assembly so it doesn't flip-flop while you're trying to remove the rings. Clamp the rod *lightly* in a vise with the bottom of the piston against the vise. If you don't have a vise, you can clamp the rod to the edge of your workbench with a C-clamp—again not too tightly.

Four items should be checked before a piston is given the OK: general damage to the dome, skirt or ring lands, ring-groove wear, piston-skirt and pin bore wear. If any one of these items proves unsatisfactory, replace the piston.

General Damage—Not so obvious damage that can render a piston useless are skirt scuffing or scoring, skirt collapse, ring-land damage and dome burning. More obvious damage can be done by something such as a valve dropping into the cylinder.

Scuffing and scoring is caused by lack of proper lubrication, excessively high operating temperatures or a bent connecting rod, all of which cause excessive pressure or temperature between the piston and cylinder wall. If there are visible scuffing or scoring marks—linear marks in the direction of piston travel—I advise replacing the piston. Scuff-marking indicates the engine was excessively overheated. If the piston has *unsymmetrical* worn surfaces on the skirt thrust faces, a twisted or bent connecting rod is the likely culprit and should be checked and corrected. An engine machine shop has the equipment to do this and it should be a normal part of their routine when rebuilding an engine to check all the connecting rods regardless of what the old pistons look like. However, if you are on a tight budget the wear pattern on the pistons will tell you what you want to know. Otherwise it's not a bad idea to have all your rods checked.

Damage to a piston's dome usually comes in the form of material being removed as a result of being overheated from detonation or preignition. The edges

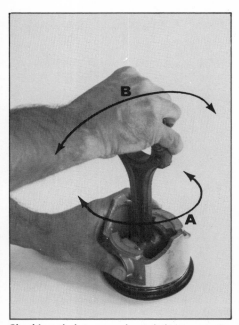
Checking pin-bore wear by twisting connecting rod in direction A, then trying to rotate it in direction B, 90° to its normal direction of rotation. If you can feel movement at the pin when holding the piston like this, pin-bore wear is excessive.

Start cleaning your pistons by removing carbon from their tops. A dull screwdriver is good for the big chunks and a wire brush will do the cleanup work, but don't touch the sides of the pistons with the brush. A piston can easily be ruined by careless use of a wire brush.

of the dome will be rounded off or there will be *porous* or *spongy-looking* areas where there was high heat concentration. To get a good look at the dome, clean off any carbon deposits. A good tool is a worn conventional screwdriver with rounded corners at its tip. The normal toolbox is usually well equiped with these. Be careful when scraping the carbon so you don't damage the piston by digging into the aluminum. Don't use a sharp-hard tool like a chisel or gasket scraper.

Detonation, or the explosion of the fuel charge, can also cause broken or distorted ring lands through impact loading. Check the top ring land for this condition. It receives the brunt of the compression loading. Consequently, if it's not damaged, the others will be OK. Broken ring lands are readily visible, but a bent one may not be, particularly without a ring in the groove to use as a reference. Reinstall a ring in the top groove and use a feeler gauge which fits snuggly between the ring and the groove. Go around the groove checking for any ring side-clearance changes which may indicate a bent ring land. The top ring land is also the one that gets the wear from an engine inhaling dirt. The wear will be on its upper surface and even all the way around.

Any of the types of damage I have just mentioned are causes for discarding a piston.

Piston-Skirt Diameter—It's micrometer time again. Mike each piston 90° to its connecting-rod pin axis in the plane of the pin. Compare this figure to what the piston mikes across the bottom of its skirts. If the skirts are not wider by at least 0.0005 in. than at the pin, the piston should be replaced because this indicates the skirt/s is partially collapsed. Skirt collapse is usually accompanied by heavy scoring or scuff marks on the skirt—sure signs that the engine was severaly over-heated. If a piston with these symptoms gets reinstalled, you'd have a noisy engine, and the piston would eventually experience total skirt failure. When an engine overheats, its pistons try to expand more than its bores. If the engine gets too hot, the piston skirts squeeze out the oil cushion between the piston and its bore. Not only does the skirt contact the bore, it is overstressed so both scuffing and skirt collapse occur. After the piston returns to its normal temperature, its bore clearance will be excessive and the engine will be noisy, particularly during initial warmup.

Piston-Pin Bore Wear—Measuring pin-bore wear requires disassembling the piston and rod assembly. A practical and easier way of doing this is to *feel* for the wear. The Nominal pin clearance is 0.0003 in. and should not exceed 0.0008 in. Excess clearance will show up when you try wiggling or rotating the connecting rod 90° to its normal direction of rotation. First clean each piston with solvent to remove any oil between the piston and the pin. Oil takes up clearance and sufficiently cushions movement to give the impression the piston-to-rod relationship is OK when it's not. Now, with the piston and rod at room temperature, if you feel movement as you try to wiggle the rod sideways, the pin bore is worn beyond the limit and the piston should be replaced. Otherwise it's OK.

Clean The Ring Grooves First—I saved this job till last because cleaning the ring grooves is a tough job and you have to do it before checking the ring grooves. So, if some or all of the pistons didn't pass your previous tests, you've avoided unnecessary work. You'll need something to clean the ring grooves without damaging them. A special tool for doing just this is called a *ring-groove cleaner*. Prices range from $10 to $30, depending on the quality of the tool. A ring-groove cleaner fits around the piston and *pilots* in the groove it is cleaning. An adjustable scraper fits in the groove and cleans carbon and sludge deposits from the groove as the cleaner is rotated around the piston. If it bothers you to make a purchase like this for a "one-time" use, use the broken end of a piston ring to clean the grooves. It takes more time and you'll have to be careful, but it can be done just as well. Be careful, no matter which method you use. Caution: Don't remove any metal or scratch the grooves, just get rid of the deposits. Be especially careful to avoid removing metal from the side surfaces of the grooves as they are the surfaces against which the rings seal.

Measuring Ring-Groove Width—Again, chances are if a ring groove is worn or damaged, it will be the top one because

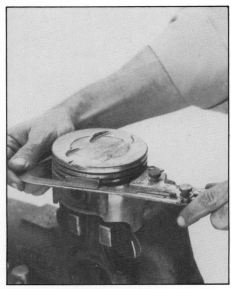
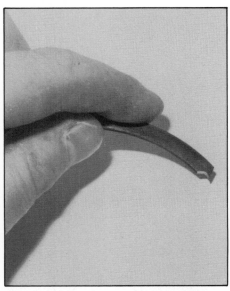

Two tools which accomplish the same thing—an official ring-groove cleaner or a broken piston ring ground like this. The ring is harder to use, but a lot cheaper. Regardless of how you clean your ring grooves, be careful not to remove any metal.

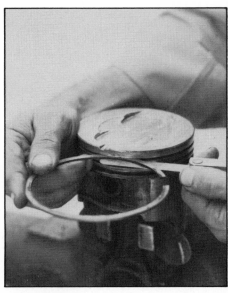

Ring and feeler gauge checking ring-groove wear. New rings should have no more than 0.006 in. side clearance.

of the higher loads. So start with it. This doesn't mean you don't have to check the others if the top one is OK. Unforeseen things can happen, so check them all. Look at the clean grooves first. Any wear will have formed a *step* on the lower portion of the ring land. The height of the step shows up as additional ring *side clearance* due to groove wear and the length of the step projecting from the back-wall of the groove represents the piston ring's *back-clearance*. Side clearance is what you're checking. It's the distance between one side of the ring and a ring land measured with a ring and a feeler gauge with the ring held against the opposite ring land. If it is too much, proper ring-to-piston sealing is not possible, plus the additional action of the ring moving up and down in the groove will accelerate wear and increase the possibility of breaking a ring land.

Compression-Ring Grooves—Compression-ring grooves are nominally 0.080-in. wide, new rings have a minimum 0.77-in. width and maximum ring side clearance is 0.006 in. You are presented with a dilemma because the 0.006-in. side clearance should be measured with a new ring in the groove. You don't want to lay out money for a new set of rings for checking your old pistons until you've checked them and have given them the OK. One way around this is to use an old ring for checking. The problem is, rings wear too, consequently more than actual side clearance will show up if ring wear is not accounted for. Just mike the old ring and *subtract* this amount from 0.077 in., the minimum width of a new compression ring. Add this figure to 0.006 for the maximum allowable *checking clearance*. I'll use the term *checking clearance* rather than *side clearance* because the gauge thickness may not be the actual side clearance with a new ring. If the old ring measured 0.076 in. it is 0.001-in. undersize. Consequently maximum checking clearance is now 0.007 in. but it will yield a 0.006-in. side clearance with a minimum-width new ring. In formula form, this looks like:

Maximum checking clearance
= 0.006 + (0.077 - 0.076) or:
Maximum checking clearance in inches
= 0.006 + (0.077 − Checking-ring width in inches)

When checking side clearance, insert the edge of the ring in the groove and insert the checking-clearance feeler gauge between the ring and the *lower side of the ring groove*. Slide the feeler and ring completely around the groove to check for any clearance variations and to verify the land is free from distortion or uneven wear. The gauge should slide the full circumference without binding. There is no need to install the ring into the groove. Check both compression-ring grooves using this method.

Oil-Ring Grooves—Nominal oil-ring grooves are 0.1885-in. wide. The real test uses a snug-fitting oil ring. Here's that old problem again. You don't yet have a new set of rings to do the checking. Fortunately oil rings and their grooves are well lubricated and aren't heavily loaded like the compression rings, particularly the top one. Consequently, their wear is minimal, so if they pass your visual inspection they should be all right. However, to be positive, measure them. To do this, stack two old compression rings together and mike their combined thickness. Now, use the stacked rings with your feeler gauge to check groove width. You don't have to install the rings in the groove. Just insert their edges in the groove with the gauge and slide them around the grooves. As an example, if the combined thickness of two rings is 0.155 in., add feeler-gauge thickness to get maximum oil-ring-groove width, or 0.191 in. minus 0.155 in. = maximum allowable checking feeler-gauge thickness of 0.036 in. If the oil-ring grooves in your pistons don't exceed this amount you can be certain the rings will fit snugly in their grooves.

Maximum checking gauge thickness in inches
= 0.191 − Combined thickness of two compression rings in inches

Checking for a bent or twisted connecting rod using a checking fixture in an engine machine shop. A bent or twisted rod can be straightened without damage.

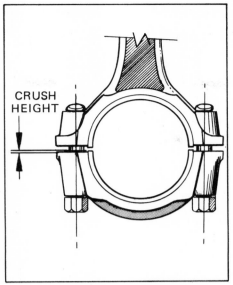

How much a bearing insert projects above its bearing-bore parting line is called *crush height* (about 0.001 in.). As a bearing cap is tightened down, the insert-half ends contact first, forcing the insert to conform to the shape of its bore and to be locked into place. This is called *bearing crush*.

Groove Inserts—If you have the unlikely circumstance that your engine block doesn't need boring, only honing to get it back into condition, and your pistons are OK for reuse except for having too much ring-groove wear, you can have your piston-ring grooves remachined wider. This makes all the ring-groove widths consistent. To compensate for the additional clearance between the rings and the grooves, ring-groove inserts are installed—they are usually 0.060-in. wide and are installed beside the rings. In terms of cost, ridge-reaming and honing your block, machining your pistons in preparation for ring-groove inserts and purchasing the rings and inserts will cost 50–60 percent of a rebore and new pistons and rings. In this case, durability is directly related to cost. You can expect approximately half the life from an engine with inserted pistons as compared to one with the complete rebore job.

CONNECTING RODS

Inspecting connecting rods involves checking three areas: out-of-round or enlarged bearing bores, twisted or bent rods and cracked rod bolts. Now's when the old bearing inserts come in handy. If a rod bearing shows uneven wear from side-to-side—opposite sides on top and bottom bearing halves—the piston on that rod has wear spots offset from its thrust face and the rod's crankshaft bearing journal was not *tapered*, the rod is bent, causing *side loading*. With these symptoms, the rod and piston assembly or assemblies should be taken to an engine machine shop for accurate checking and straightening if necessary. Bearing condition also tells you if the "big end" of the rod needs *reconditioning*. When a rod is reconditioned the bearing bore is checked with a special dial indicator to determine its shape—round, out-of-round or oversize. This is basically the same thing you did with the crankshaft bearing journals, however now the check is made of the bore in which the bearings are retained. With an out-of-round bore, its bearing inserts will assume the same irregularity, causing uneven load distribution between the bearing and its journal, resulting in uneven and accelerated bearing wear. Consequently, the bore must be reconditioned.

A bearing with a too-large bore is even worse. Oversized bearing bores will let bearings move in the bore—which is not supposed to happen. The reason is the insert halves are not sufficiently *crushed*. Crush occurs when a bearing is *forced* into its bore. This is accomplished by the circumference of the outside diameter—half of a diameter for each insert half—of the insert being more than that of the bearing bore. When a bearing is placed in its bore its ends project slightly above the bearing housing's parting surfaces. Consequently, when two bearing inserts are installed in their bore, the ends of the inserts butt. As the bearing cap is tightened, the two circumferences must become equal. The bearing shell gives—is crushed— causing the bearing to assume the shape of its bore and to be preloaded, or fitted very tightly in its bore. This tight fit and the tooth, or machining marks in the bearing bore, combine to prevent the bearing inserts from *spinning* in their bores. The reverse happens when the force at the bearing journal which tries to rotate the bearing overcomes the force between the bearing and its housing, or bore that is resisting this force. If bearing bore is too large the bearing may move in the bore, or worse yet, spin.

To determine if a bearing has been moving in its bore, look at its backside. Shiny spots on the back of the shell indicate movement. If this happened, either the bearing bore is too large or the bearing-to-journal clearance was insufficient. If you discover any shiny spots the rod/s need to be checked with a dial-indicator that's part of an engine builder's connecting-rod reconditioning hone. Recondition the rod/s as necessary.

When a connecting rod is reconditioned, some material is precision ground from the bearing cap mating surfaces. Next the cap is reinstalled on the rod, bolts are torqued to spec and the bore is honed to the correct diameter. Honing corrects the bearing-bore diameter and concentricity and also restores the *tooth*, or surface of the bore which *grabs* the inserts to prevent them from spinning.

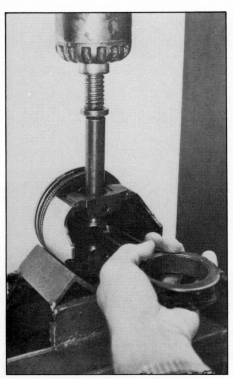
If your pistons have to be replaced this is the first thing you'll have to do—disassemble them. A press is required.

Connecting-rod bearing bore being checked and honed, or reconditioned. If the bearing bore isn't nearly perfectly round, some material is precision ground from the bearing cap parting line. The cap is then reassembled to the rod and the nuts torqued to spec. The bearing bore is then honed to its correct diameter.

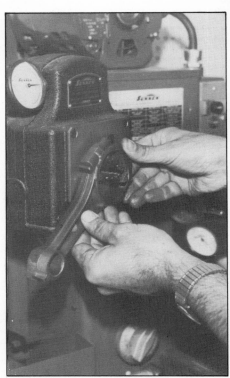

Rod Bolts—Make sure you inspect the rod bolts very closely *before* attempting any other reconditioning of the rods. Don't hesitate to use a magnifying glass as you give them the old "eagle-eye." Replace any cracked bolts—a cracked rod bolt will eventually break and may totally destroy an engine.

You have to remove the bolts to check them. To remove a rod bolt, clamp the big end in a vise between two blocks of wood and drive the bolt straight out. The new bolt can be installed by tightening it after it is loosely installed in the rod with the cap. If you replace one or both rod bolts, *the connecting rod must be reconditioned* or honed because the bolts locate the cap in relation to the rod. Consequently, changing a bolt may move the cap position very slightly, throwing the bearing bore out-of-round.

One last note concerning connecting rods. Rod-bearing bores are originally machined with their caps torqued in place, so they must be checked the same way. When you deliver them to a machine shop for checking and reconditioning after you've inspected the rod bolts, make sure the correct cap is installed in each rod. Torque the nuts to specification before making the delivery. Torque the 5/16-in. rod-bolt nuts of the 221, 260, 289 and 302 engines to 19–24 ft. lbs. Torque the larger 3/8-in. rod bolts and nuts of the HP289, Boss 302 and 351W engines to 40–45 ft. lbs.

DISASSEMBLING AND ASSEMBLING PISTONS AND CONNECTING RODS

Disassembling or assembling connecting rods and pistons is a job for an expert with special equipment. If you are replacing your pistons, the rods and pistons must obviously part company. Only one method can be used to do this correctly. The connecting-rod wrist pin must be pressed, **not driven**, out of the small end of the rod in which it is retained by a 0.001–0.0015-in. interference fit between the pin and the rod. A press with mandrels to back up the piston and bear on the pin is required.

Assembling a connecting rod and piston can be accomplished by one of two methods. The first is done by reversing the disassembly process, however it must be done with considerably more care. The other method, and the one I prefer, is the heating method. This is done by simply heating the small end of the connecting rod with a torch so it expands, allowing the pin to slide into place without need for a press. The rod should not be heated excessively because this could damage the rod. Pin and piston have to be fitted very quickly before the rod and pin approach the same temperature. Otherwise the pin is trapped by the interference fit before it's in position, requiring a press to complete the installation.

If you are assembling your own pistons and rods, be aware of the piston's location relative to its connecting rod. As installed in the engine, all pistons have a notch at the front edge of their domes or an arrow stamped in the dome, both of *which must be pointing to the front of the engine* as installed. Also, the *connecting-rod numbers must point toward their cylinder bank*. Rods installed in the right bank must point to the right and those in the left bank must point to the left. So, putting them together, you'll have piston and rod numbers 1, 2, 3 and 4 with their *notches or arrows pointing forward and the rod numbers to the right*. Piston and rod numbers 5, 6, 7 and 8 have their *piston notches or arrows pointing forward*, and their rod numbers to the left.

TIMING CHAIN AND SPROCKETS

I previously discussed timing chain wear and how to determine if a chain and

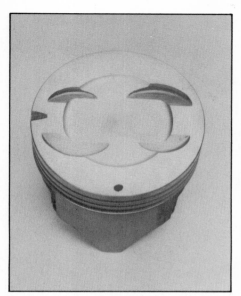
Notch, which could also be an arrow, indicates the position of a piston in its bore. Notch or arrow should point to the engine front. When assembling connecting rods and pistons, rod numbers 1, 2, 3 and 4 must be to the right of the piston (opposite the dot) and numbers 5, 6, 7 and 8 to the left (same side as the dot).

Piston and rod assembling being done by enlarging the wrist-pin bores by heating with a propane torch. The pin slips right into place, but you've got to be fast. Rod and wrist-pin temperatures converge very quickly, causing the two to lock together within a couple of seconds. If the pin is not located properly when this happens it'll have to be pressed into place.

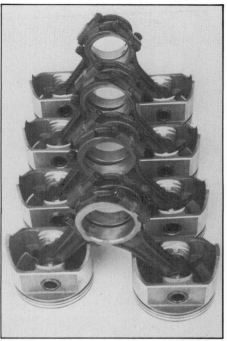
Pistons and rods assembled and ready for installation and positioned as they would be as installed in their engine. Looking from the front of the engine, notches or arrows should be pointing at you with the rod numbers on the side of their respective row of pistons.

its sprockets need replacing. However here's some more information about what replacement parts you should consider. As I said, if your engine is one of the hotter-operating 1973 and later engines—and you live in a hot climate, you should change to the cast-iron type cam sprocket if your engine was originally equipped with the nylon type. This avoids the problem of cracked and broken sprocket teeth, but don't expect better sprocket or chain wear. There's negligible difference.

Two original-equipment chains have been used. The original C20Z-6268-A must be used with 1/2-in.-wide sprockets. The one originally intended for the HP289 and carried over for use with the nylon cam sprocket, C30Z-6268-A, is for use with 13/32-in.-wide sprockets. Therefore, if you plan on changing components from what your engine was originally equipped with, use the nearby chart to determine which parts are compatible. One popular change is to replace the nylon cam sprocket with the cast-iron one—but which one? If you use the earlier standard C30Z-6256-C sprocket, you'll have to change the crankshaft sprocket *and* chain because they are wider than the nylon-sprocket-matched chain and crankshaft sprocket. Because the HP289 cam sprocket is compatible with the narrow chain, you can install it by replacing worn-out parts *and the thrust plate*. It must be changed because it's thicker than those used on all other small blocks. For example, HP289 thrust plate—C30Z-6269-A is 0.375-inch thick; others vary between 0.22- and 0.25-inch thick. Because the back of the cam-timing gear bears against the front of the thrust plate and the front-cam-bearing journal bears against the back of the thrust plate, the thrust plate controls camshaft end-play. Consequently, if the thrust plate is too thick, there will be no end-play and the cam will be locked up. On the other hand, if it is too thin, excessive cam end-play could cause serious cam and lifter damage. So, if you install any parts other than those originally in your engine, make sure they are *all compatible*.

Replacement-part manufacturers have simplified the job of installing either type of cam sprocket you prefer. TRW offers both nylon and cast-iron cam timing gears for all small blocks except 221 and 260, for which only cast-iron gears are offered. This makes it convenient because the gear specified for a specific engine can be installed on that engine without any other hardware changes. In other words, chains and gears are manufactured to accommodate the hardware originally found on the engines for which they are specified.

Heavy-Duty or Performance Applications—In addition to the OEM and OEM-replacement sprocket and chain sets, high-performance and heavy-duty sprockets and chains are available. Consider these if your engine is going to be used for extremely hard service. They stretch less and are relatively unaffected by high engine-operating temperatures. The chain I am referring to is double-roller type with matching sprockets, similar in design to bicycle chains except with two rows of rollers. These are available from Cloyes Gear and Products, Inc., 4520 Beidler Road, Willoughby, Ohio 44094 and TRW, 8001 E. Pleasant Valley Road, Cleveland, Ohio 44131.

Camshaft End-Play—I could wait until the engine-buildup chapter to talk about camshaft end-play, but it's a good idea to find out if you need a part now instead of waiting for a long holiday or late Saturday night when everything is closed and you are in the middle of assembling your engine.

A camshaft doesn't have to be installed in an engine to check its end-play. Just install the sprocket on the cam with the thrust plate in its normal position—between the sprocket and the front bearing journal. These should be the parts you intend to use in your engine: cam, sprocket and thrust plate. Sprocket and cam thrust faces wear. You won't get a true reading when checking end-play if you check with old parts and then install new ones. Now, use your feeler gauges to check the clearance between the thrust plate and the front cam-bearing-journal thrust face. The maximum-thickness feeler gauge represents camshaft end-play. End-play should be 0.001–0.007 in. If it exceeds 0.007-in. replace the thrust plate.

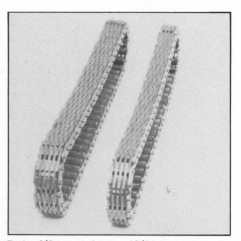
Early 1/2-in.- and later 13/32-in.-wide timing chains. They must be used with the correct width sprockets.

Nylon/aluminum camshaft sprocket has built-in spacer (arrow) as do aftermarket cast-iron sprockets. OEM cast-iron sprockets, the ones purchased from a Ford dealer, require a separate C-shaped spacer.

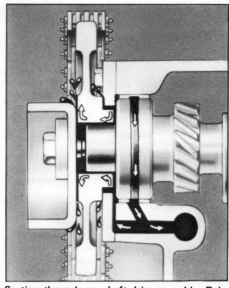
Section through camshaft drive assembly. Relationship between thrust plate and sprocket spacer must be correct or the cam will be locked up—or excessive end play results. Note how the fuel-pump cam retains the sprocket to the camshaft. Drawing courtesy Ford.

TIMING CHAIN & SPROCKET APPLICATIONS

ENGINE	221-260-289	221-260-289	260-289-302-351W	302-351W	HP289	HP289	Boss 302
YEAR	1962 to 9-2-63*	9-2-63 to 1965*	1965 to 5-2-72 ●	5-2-72 to Present ●	1963 to 1965*	1965 to 1967 ●	1969 to 1970
CAMSHAFT SPROCKET							
Cast Iron	C3OZ-6256-C	C3OZ-6256-C			C3OZ-6256-A		
Nylon			C5OZ-6256-B	D3AZ-6256-B		C5OZ-6256-B	C5OZ-6256-B
CHAIN	C2OZ-6268-A	C2OZ-6268-A	C3OZ-6268-A	C3OZ-6268-A	C3OZ-6268-A	C3OZ-6268-A	C3OZ-6268-A
CRANKSHAFT SPROCKET	C2OZ-6306-A	C2OZ-6306-A	D5OZ-6306-A	D5OZ-6306-A	C3OZ-6306-A	C3OZ-6306-A	D5OZ-6306-A
THRUST PLATE	C2OZ-6269-A	C3OZ-6269-B	C9OZ-6269-A	C9OZ-6269-A	C3OZ-6269-A	C9OZ-6269-A	C9OZ-6269-A
THRUST PLATE BOLT (2 Req'd.)	1/4-20 x 5/8 spl. head.	1/4-20 x 5/8	1/4-20 x 5/8	1/4-20 x 5/8	1/4-20 x 5/8	1/4-20 x 5/8	1/4-20 x 5/8
SPACER	C2OZ-6265-A	C2OZ-6265-A	NOT REQ'D.	NOT REQ'D.	C3OZ-6265-A	NOT REQ'D.	NOT REQ'D.

*Before Change L-7
● After Change L-7

Chain no. C2OZ-6268-A is for use with 1/2-inch-wide sprockets and chain no. C3OZ-6268-A is used with 13/32-inch-wide sprockets.

Use this chart to match up camshaft-drive components. Numbers shown are current service-part replacement numbers.

If your engine is going to be subjected to severe service, consider using a timing chain and sprocket setup like this. It's Cloyes True Roller® which uses a roller-chain design rather than the conventional silent chain. Note the two extra keyway slots in the crankshaft sprocket for advancing or retarding camshaft timing.

OIL PUMP AND DRIVE SHAFT

Lubrication is a major key to an engine's durability, and the oil pump is the heart of the lubrication system so don't take any short cuts here. Check your oil pump for any internal damage such as grooving or scoring of the rotors or the housing, and check clearances before reusing it. While it's one of the most durable components in your engine, it can wear past the point of being reusable. The hex-shaft that drives it should be replaced. *Don't even consider* reusing your old shaft in your new engine. The hex edges wear and will more than likely fail later. Remember, the distributor continues to turn, allowing the engine to run, but a failed shaft means the engine is not being lubricated. It doesn't take much imagination to visualize the resulting damage.

The oil-pump drive shaft for the 302 engines has a 1/4-in. hex 7.44-in. long. The 351W shaft hex is 5/16-in. 8-3/16-in. long. Part numbers are D8AZ-6A618-A for the 221–302 and C9OZ-6A618-A for the 351W.

Oil-Pump Damage—Before I get much further into the oil-pump subject, let's talk about durability. An oil pump is so over-lubricated that it's not going to wear much. Damage occurs when metal or dirt particles pass through the pump. This happens when some other component is chewed up—such as a cam lobe and lifter—or the oil filter clogs up and oil *bypasses* the filter, resulting in large dirt particles circulating through the engine and the oil pump. This can cause severe scoring and grooving of the oil pump rotors and body as well as bearing journals on the crankshaft and cam.

If your engine exhibits either condition, you may have already found oil-pump damage. Clues will have been given by the components you've already inspected and reconditioned, or replaced. Take this into account during your oil-pump inspection. For example, an engine which had heavily scuffed pistons and scored cylinder walls, wiped out camshaft lobes and lifters or deeply grooved crankshaft bearing journals will have had large amounts of metal and dirt particles circulated through the oil pump.

Oil-Pump Inspection—Test your oil pump by immersing the pickup in a pan of clean solvent, then turning the rotors with your fingers using the hex shaft. After two or three complete turns solvent should *gush* out of the pump's pressure port—the hole in the center of the pump-mounting flange. If the pump performs with solvent its internal clearances are not excessive, plus it shows the rotors turn freely.

Next inspect the pump internally. Remove the four 1/4-in. bolts and the cover from the pump body. Be careful the rotors don't fall out. Use your feeler gauges to determine the maximum clearance between the oil-pump housing and the outer rotor. It should not exceed 0.013 inch. Check the rotor end-play by laying the cover plate over half the housing and measuring clearance between the plate and both the inner rotor and the outer rotor with your feeler gauges. Maximum clearance is 0.004 inch. Next remove the rotors. When turning the pump over, be ready to catch the rotors as they fall out. Be careful not to bang them around. Also, keep them oiled to prevent rusting. Look at both rotors and the interior of the pump housing for signs of scoring, pitting or deep grooving, particularly between the inlet and outlet ports in the housing. A sure sign that the pump ingested foreign particles is you'll find some imbedded in the tips of the inner rotor—another condition requiring replacement. If the rotors are damaged they can be replaced for about a third of the cost of a new pump. Rotors have to be replaced in pairs and Ford offers rotor kits for the 221–302 and the 351W. Kit numbers are B8A-6608-A and C8SZ-6608-A, respectively. Just make sure if you decide to do this that the surfaces in the pump body aren't damaged, otherwise little will be gained by installing new rotors.

Pressure-Relief Valve—The major concern with your oil-pressure relief valve is that the valve and spring operates freely. To check for this, insert a small screwdriver into the *pressure port* and move the relief valve. If it moves freely, then it's OK.

Assemble the Pump—Make sure the parts are clean. Install the rotors, then coat them liberally with oil. Make sure the outer rotor is installed correctly. If an *indent* mark is not showing similar to the one on the end of the inner rotor, turn the outer rotor over to expose the mark. Install the cover plate and torque the bolts 6–9 ft. lbs.

Before leaving the oil-pump subject I'll tell a little story to give you ammunition for when you are trying to convince anyone of the dangers of driving a vehicle or operating any engine with little or no oil pressure. Any time an oil-pressure light comes on or a gauge reads low pressure, an engine should be shut off immediately, not after you make it to the next gas station, another block, mile or whatever, but right **now**. A fellow I know had the engine in his pickup seize due to oil-pressure loss when the pump drive shaft failed. When I questioned him about whether or not he saw the *idiot light* he said, " Yes, but it wasn't bright red—just pink!" Using this reasoning, he tried to drive the last 10 miles home but only made it five miles before the engine seized. The repair job came to approximately $500, or $100 per mile for the last five miles *after* he saw the light.

  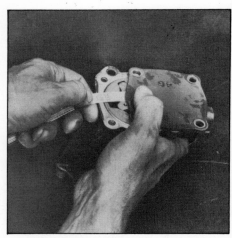

Disassemble your oil pump and check for scoring between the inner and outer rotors and the pump housing. Also check clearances. If they are excessive your pump won't be able to produce sufficient pressure, consequently it should be replaced.

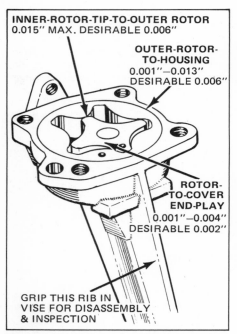

Here are the clearances you should be checking for. Use your feeler gauges and the cover plate as shown in the nearby photograph to check rotor end-play.

This illustrates why you should always keep your eyes open. Crack (arrow) didn't appear until I installed the pickup after the pump was cleaned, inspected and reassembled. Tightening the bolt opened up the crack. If I hadn't seen it, the mounting flange could have broken, letting the pump suck air rather than oil. Air doesn't lubricate an engine very well.

New and used oil-pump drive shafts. Don't try to save money by using your old one.

Head Reconditioning and Assembly 6

1969 Boss 302 Mustang engine with its unique cylinder heads. The major objective behind this engine was to provide a practical production engine that would be the basis for the strongest and most powerful 5-liter (302 CID) engine for competition in the SCCA Trans Am racing series—it was and they won. Photo courtesy Ford.

I reserved cylinder-head disassembly and inspection till now to go with reconditioning because cylinder-head work must be done in an orderly manner. Once you start tearing your heads down, you should continue with them until the job is finished. This will greatly reduce the possibility of losing or mixing up pieces. Head reconditioning is a precise, tedious and dirty job which requires special tools or equipment. The best policy is to limit your head work to removing and installing them. You may do more harm than good if you attempt to do more. Leave the precision work to the specialist. It'll cost you no more for the specialist to remove the valves. However, if you have access to some or all of the equipment, it's not going to cost you any more to disassemble the heads. By doing this you'll be able to inspect the parts so you'll know what must be done to put your heads back in shape.

Rail-Rocker-to-Spring-Retainer Clearance—If your engine is equipped with the rail-type rocker arms, which were standard for all small-blocks from late 66 to 1977, check the clearance between the rails on each rocker arm and its spring retainer. This dimension should not be less than 0.0625 in. (1/16 in.)—original clearance is 0.085 in. A not-too-common occurrence is for the valve-stem tip and rocker arm to wear to the point that this clearance is completely gone. As a result the rocker-arm rails begin to bear on the spring retainer, eventually causing the retainers to come loose. I mentioned this earlier in the teardown chapter, but because of its importance I'm repeating it. A dropped valve can literally destroy an engine, so be careful when checking rocker-arm-to-spring-retainer relationships.

Locate each rocker arm in its normal position on its valve-stem tip and measure the clearance between the rails and the spring retainer. A six-inch scale is handy for doing this, or use a 0.0625 in. thickness of steel or aluminum sheet as a feeler gauge. Discard any *valve and rocker-arm combination* under the 1/16-in. limit. The reason for replacing the rocker arm too is that excessive contact wear on one component means the other component is also too far gone—it will have also worn in an erratic pattern. So, if the old rocker arm were to be used with a new valve, there would be excessive valve-tip loading causing accelerated valve-tip and rocker-arm wear.

TEARDOWN AND CLEANUP

Keep the Rocker Arms and Their Fulcrums Together—When removing the rocker arms, their fulcrums and adjusting nuts remember to keep each fulcrum with its rocker arm. String them on a piece of wire, less the ones you've found to be excessively worn. Just like camshaft lobes and lifters, rocker arms and their fulcrums *wear in* together. Consequently, mixing them up will result in galling.

Check Rocker-Arm Adjusting-Nut Torque—Two methods have been used for small-block Ford valve adjustment. Pre 68-1/2 valves were adjusted by moving the rocker-arm nut vertically on the rocker-arm-stud thread. After 68-1/2 the thread on the stud was reduced in size from 3/8-24 (3/8-inch diameter and 24 threads-per-inch) to 5/16-24, but the end pressed into the cylinder head was maintained at 3/8-in. This was designated as *Change L4*. Because the thread diameter was reduced but the unthreaded end diameter maintained, a shoulder was created at the bottom of the threads. Consequently, the rocker-arm fulcrum is clamped between this shoulder and the attaching nut—it's not adjustable anymore, it is a *positive-stop* setup. Valves are adjusted in the after-68-1/2 engines by installing longer or shorter pushrods.

Valve adjustment in the pre-68-1/2 is maintained by the locking capability of the adjusting nut. These nuts are of the *locking* variety, not the *free-running* type. If you look at the end of the nut, you'll see a definite triangular shape rather than a round one. This deforms the threads so there is friction between the nut and the rocker-arm stud which causes some amount of *breakaway* torque, or torque to start the nut turning. This breakaway torque should be a minimum of 5 ft. lbs. if your engine is equipped with hydraulic lifters, or 7 ft. lbs. for the mechanical-liftered HP289. The load and vibration transferred to valve train by a solid lifter will cause the nut to lose adjustment more than a hydraulic lifter. Although a solid-lifter valve-train adjustment is immediately affected by any adjustment or nut movement, hydraulic lifters compensate for changes within limits.

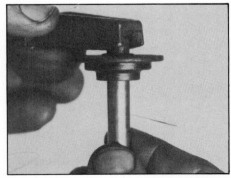

If the tip of a valve is worn so the rocker-arm rails are less than 1/16-in. from the valve retainer after its tip has been reconditioned, the valve should be replaced.

75

Real trouble about to occur: Shiny spots on rocker-arm rails and spring retainer indicate rocker arm is beginning to push on retainer rather than the valve tip. This will eventually loosen the keepers, letting the valve drop into the cylinder. Valve and rocker arm must be replaced.

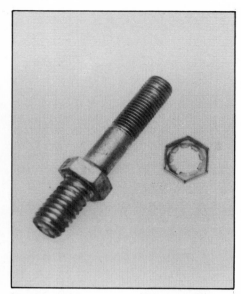
Screw-in stud is standard equipment on HP289 and Boss 302 engines. If you have the HP289, this Palnut® will increase time between rocker-arm adjustments. Boss 302 is already equipped with a jam nut as shown by this Ford drawing of a Boss 302 rocker-arm-and-pivot assembly.

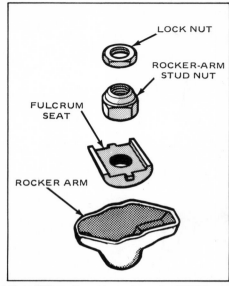

When removing your rocker arms, take time to check nut breakaway torque. For the positive-stop type setup used after 68-1/2, be particularly careful to check them for stress cracks on their undersides radiating out from the threads. A cracked nut may split in two. Therefore, if a nut is below minimum torque or it's cracked, replace it with an OEM or OEM replacement nut. Part numbers are C2DZ-6A529-A for the pre-68-1/2 engines and D0AZ-6A529-A for the later ones. *Don't substitute different style nuts. Jam nuts can be used.* If your rocker-arm nuts are in good shape, but some are below minimum torque, rather than purchasing new locknuts, purchase 16 3/8-24 jam nuts. Install them on top of the regular adjusting nuts. This works especially well on the HP289. I'll say more about this later. The Boss 302 with its 7/16-20 studs was originally equipped with jam nuts.

Removing the Valves and Springs—Now, you'll need the first piece of equipment not normally found in the standard tool chest—a *valve-spring compressor.* A valve-spring compressor is a specialized spring compressor suited to cylinder-head work. They are made specifically for compressing valve springs so the keepers, locks, collets, keys or whatever you want to call them can be removed or installed so the valve spring, spring retainer, valve and seal can be removed or installed. There are two types of spring compressors, the *C-type* and the *fork-type.* If you have a choice, pick the C-type. The fork-type is used for working on heads installed on the engine, so forget it because of its difficulty to use. The C-type compressor straddles the head and pushes on the valve head and on the spring simultaneously. This forces the spring into a compressed position while the valve stem projects out of the retainer so the keepers can be removed or installed.

Loosening Sticky Keepers—When compressing the valve springs, you'll probably encounter sticky keepers. The trick here is to place the head right-side-up on your bench or a block of wood, and using a 9/16-in., 3/8-in. drive deep socket as a driver, place the socket squarely over the valve tip and on the retainer. A sharp rap on the socket with your soft-faced mallet will not only break loose the spring-retainer keepers, it may pop them out as well, releasing the retainer and spring. If you're lucky you won't have to use the spring compressor. A small block of wood, about 2 in. x 3 in. x 3/4 in., under the valve head in the combustion chamber will keep the valve from moving and the keepers will literally "fly off." If this method doesn't work 100% you have to revert to the spring compressor and finish removing the keepers.

With the keepers out of the way and in a container where they won't get lost, you can release the pressure on the spring and remove the retainer and spring—if you're using a compressor. Slide the valve out of its seal and valve guide. Store the valves in order. Throw the seals away. They will be replaced just like the gaskets. To keep the valves in order, use anything from a cardboard box to a piece of lath with 16 holes drilled in it. '68-1/2 through '71 engines will have loose *valve-stem-tip caps* on the exhaust valves. Tape them to their valves so they won't get lost.

In addition to the stem-tip caps, spring retainers were changed to a two-piece assembly in '68-1/2 as part of *Change L4.* Rather than the one-piece machined retainer, the two-piece retainer consists of a spring retainer and a *spring-retainer sleeve* which fits between the retainer and the keepers. The change does not affect interchangeability, you can use one retainer in place of the other.

Exhaust-Valve-Stem-Tip Caps—Exhaust-valve caps are essentially hardened-steel hats fitted on the end of the exhaust valves between each valve-stem tip and rocker arm. The caps minimize valve-stem-tip wear caused by rail-type rocker arms. Beginning in 1969, all 302s in taxis, police cruisers and Broncos use caps permanently fixed to their valves. After 1972, all 302s and 351Ws use the fixed-type caps.

Removing valve and spring assemblies from a cylinder head requires a spring compressor. A tap on the spring retainer loosens the keepers so the spring can be compressed and the keepers removed.

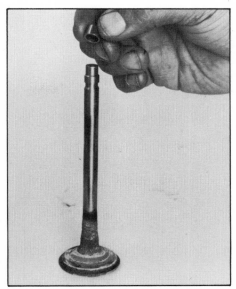
If your valves are equipped with removable tip caps, don't lose them

Freeze plugs are removed from heads like they were from the block—knock them in, then pry them out.

Putting the good old gasket scraper to work.

Change L4—To sum up Change L4, the following parts were affected: Rocker-arm stud threads and nuts were reduced from 3/8-24 to 5/16-24 to stop, or locate the rocker-arm fulcrums positively. This requires different pushrods for valve adjustment—6.905-in. standard-length pushrods with 0.060-in. longer or shorter pushrods for adjustment. Single-piece spring retainers were changed to two-piece. Finally, hardened steel caps are used on the exhaust-valve tips only. Change L4 was instituted in mid-1968 in the 302 engines and these changes were incorporated in the *new* 351W engine when introduced in 1969.

Knock Out the Freeze Plugs—With the valves and related components gone the only thing left to remove from the heads are the freeze plugs—one at each end. Remove them using the same method you used when removing the plugs from the block. Knock them out—actually in—and then pry them out with pliers. Now your heads are bare with the exception of the rocker-arm studs, grease, dirt and old gasket material.

CLEANUP AND INSPECTION
Scrape, Scrape, Scrape—Time for the old gasket scraper and elbow grease again. As you probably found when scraping the deck surface of your engine block, head gaskets seem to grow to the metal. If you had your heads hot-tanked the job will be easier and a lot cleaner. Clean all gasket surfaces, the head-gasket, intake-manifold and rocker-arm-cover-sealing surfaces of all old gasket material. Using a round file, clean the head-to-block water passages of all deposits.

It's time again to *misuse* a screwdriver or a small chisel. Scrape the carbon deposits from the combustion chambers and intake and exhaust ports. After the heavy work is done, finish the job with a wire brush. This will put a light polish on the surfaces so you can spot any cracks which may have developed in the combustion chambers, particularly around the valve-seat area. If you do find cracks in any of these areas, consider the head junk and replace it even though it could be repaired by welding. Chances are the repair cost would exceed the replacement cost.

Head-Surface Flatness—Because cylinder heads are subjected to extreme heat, compression and combustion loads and are structurally weak compared to an engine block, they are susceptible to warpage. Excessive mismatch between the block and head surfaces causes combustion or coolant leakages, consequently those surfaces *must be flat*. Check the head-gasket surfaces with a *straight-edge* and feeler gauges. Set the straight-edge lengthwise and diagonally across the head in both directions. Measure clearances which

Get those gasket surfaces clean.

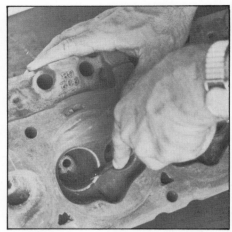

Make carbon removal part of your cylinder-head cleanup. I'm using a screwdriver for the heavy work, and will finish up with a wire brush.

INTAKE-MANIFOLD VERSUS CYLINDER-HEAD MILLING REQUIREMENTS (INCHES)

Removed from Cylinder Heads	Removed from Manifold Bottom	*Remove from Manifold Sides
0.020	0.028	0.020
0.025	0.035	0.025
0.030	0.042	0.030
0.035	0.049	0.035
0.040	0.056	0.040
0.045	0.063	0.045
0.050	0.070	0.050
0.055	0.078	0.055
0.060	0.084	0.060

*Double amount when milling one side of manifold.

Resurfacing a warped cylinder head. Use accompanying chart to determine intake-manifold milling requirements if more than 0.020 in. is removed from heads.

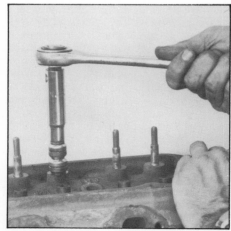

Press-in stud being removed by shimming up from stud boss with rocker-arm fulcrums. Free-running nut is tightened down on fulcrums using a deep socket and ratchet to draw stud out. Add shims as required to keep from running out of threads.

New stud ready for installation. Jam a couple of nuts on the stud with the top nut even with the end of the stud, then drive it into place using a soft hammer. End of stud should be even with the others.

appear between the head and the straight-edge with your feeler gauges. The maximum allowable variation in the gasket surfaces is 0.003 inch in any six inches of length and 0.006 inch for the overall length of the head for engines with 10:1 compression or more; 0.007 inch for those with lower compression ratios. Limit warpage to half of the figures I just quoted if you plan on using *shim-type* gaskets as opposed to the more compliant composition gaskets.

Cylinder-Head Milling—Whether one or both of your heads need to be resurfaced do them both. You don't want to reduce the combustion chamber volumes in one head and not the other. The result would be an increase in the compression ratio of one bank of cylinders causing more compression on one side of your engine.

Although more material can be removed from a cylinder head, I suggest you limit the amount to 0.010 inch. If necessary, a small-block Ford head can be milled 0.040 inch, however the problem with going this far is you'll boost your engine's compression ratio substantially which is not a good idea with currently available fuels. This is particularly critical if your engine is one of the pre-1972 premium-fuel engines. Mill the heads on one of these engines and you will probably end up with an engine which can't be operated without detonating.

Another consideration when milling your cylinder heads is the intake manifold. It must be milled too if more than 0.020 inch is removed from the heads because the intake ports, water passages and intake-manifold bolt holes move down and toward the center of the engine. Consequently, they will not line up with those in the intake manifold. Therefore, the manifold's bottom and cylinder-head gasket surfaces or sides must also be milled. Less than 0.020 inch removed from the heads is not significant enough to cause head and manifold alignment problems. Any more than this amount requires removing material from the manifold also. Refer to the nearby chart to determine manifold machining requirements with respect to cylinder-head milling.

Rocker-Arm-Stud Replacement—Three defects require the replacement of a rocker-arm stud: damaged threads, a broken stud and a loose stud. Damaged threads range from those that are stripped, to those on which the adjusting nut doesn't have the minimum breakaway torque, regardless of whether the nut used for checking is new or not, and finally to one that has a notch worn in it from the rocker arm rolling over on its side and contacting the stud with the inside edge of its clearance hole. The last one may eventually result in a broken stud if not replaced. As for a broken stud, chances are you won't find any unless it was the "straw that broke the camel's back," or it was what made you decide to rebuild your engine. The final problem is a loose stud. Just as with a broken stud, chances are your engine won't have one because it will have pulled out of the head while your engine was still running, consequently it would have been replaced. But, again, it may have been the "final straw."

All small-block Fords, with exception of the HP 289, Boss 302 and those using pedestal-type pivots, use pressed-in rocker-arm studs. The studs are pressed into holes smaller in diameter than the studs. The interference fit prevents the stud from coming out. The HP 289 uses screw-in-type studs because the press-in type would eventually pull out of the head due to higher loads imposed by high-rate valve springs and solid, or mechanical lifters. The cylinder heads are drilled and tapped to accept a threaded stud. Stud C30Z-6A527-B has a 7/16-14 thread at the cylinder-head end, a hex for locating and tightening the stud in the head and the pre-68-1/2 3/8-24 threads at the rocker-arm end.

Rocker-Arm-Stud Removal—The procedure required to remove a damaged press-in type stud depends on how the stud is damaged. A special stud remover locates around the stud and seats on the cylinder head as a nut is run down on the stud thread and bottoms against the puller. The puller has a ball-bearing seat so there will be little friction at the puller as the nut pulls the stud from the head as it is threaded down on the stud. You can pull a stud without the use of a special puller by using one of your sockets and a stack of washers between the socket and the bottom of the nut. To reduce friction at the nut and washers, grease them lightly. As the stud moves out of the head, you'll have to add additional washers under the nut when it begins to run out of thread.

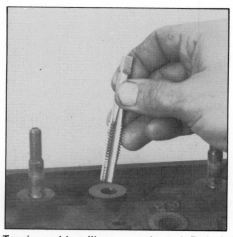 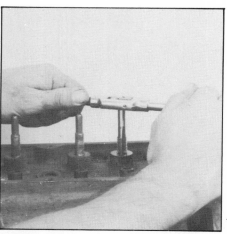 

Tapping and installing a screw-in stud. End of the 7/16-14 tap has been ground on the end for piloting in the original hole so the stud will be straight when installed. With some locking sealer on the threads, screw in the stud, using a couple of nuts jammed together. Check the position of the stud with a straight-edge across the tops of the studs or with a ruler by measuring up from the stud boss or stand. This sequence shows installing an after-market stud. Machining is required to install an HP289 screw-in stud.

All this sounds relatively easy, and it is, until you consider the type of damage to the stud. For example, if by chance the stud is loose, the job will be easy. On the other hand, if it is broken off, it won't be easy and the degree of difficulty depends where it's broken. Normally a stud breaks off at the end of the threads. Consequently, there are no threads to pull the stud, so you'll have to put threads on the unthreaded portion with a thread die. Use a 3/8-24 die and put as many threads on the remainder of the stud as you can. The pulling technique will be the same except there probably won't be room for a socket, so just use a stack of washers to pull against. The problem is compounded when a stud breaks off farther down or flush with the head where you can't get enough, or any threads on the stud. You'll then have to drill the stud out. Start by center-punching the *center of the stud* so you get a good start with the drill and won't end up drilling off into the cylinder head. For the same reason, *drill a pilot hole* in the stud—an 1/8-in. diameter drill will do—while being careful to keep the drill square with the stud. Increase the size of the drills progressively until you break out the side of the stud, or have come very close to it. When this happens, the interference fit between the stud and the head will be relieved, leaving the stud free to be removed. This can be done with a bolt extractor. Screw it in the center of the stud and pull the stud out by clamping on the end of the extractor with a pair of Vise-Grip® pliers.

For a stud that's stripped, it might have a portion of good threads that can be used for pulling. If this is the case, you won't have as much thread to work with so you'll have to pull the stud a little at a time in between adding washers. If most of the thread is damaged you'll then have to cut new threads on the unthreaded portion of the stud, as with a broken stud, and pull it out using the same method.

Removing a *threaded* stud is relatively easy. All you have to do is unscrew it unless it's broken off below the hex portion of the stud. You'll then have to drill the remainder of the stud out like is done with the press-in type, however, be careful not to get into the threads. When the hole in the center of the stud is big enough to accept a bolt extractor, install one and back the stud out.

Rocker-Arm-Stud Installation—If the stud you're replacing was broken or had damaged threads, you can replace it with a standard diameter stud if it's the press-in type. There's a special tool made just for this purpose, but if you stack three free-running nuts, as opposed to the locking type, on the end of the stud with the last nut projecting slightly above the end of the stud and tighten the nuts against one another, you can drive the stud in with a brass mallet. *Don't hammer on the end of a stud to install it,* otherwise you'll ruin the threads and the stud. Before driving the stud in, wipe some moly grease on the pilot end of the stud. This will let the stud install easier without galling it or the hole in the head. Start the stud by lining it up with the hole as best as you can. Tap it in a little at a time until 1-3/4-inches is left projecting out of the head. The best way of checking this is to scribe a mark on the stud 1-3/4-inches down from the threaded end. Stop driving the stud in when this mark lines up with the stud boss and you've got it.

Oversize Studs—Stud installation becomes complicated when you have to replace one that was loose. Because it was loose, the hole in the head will be enlarged. Consequently, if you attempt to replace the loose one with the same size stud, it will also be loose. To do the job right you have to install an oversize stud if you decide to stay with the press-in type. I say this because there's an easier way of replacing rocker-arm studs which I'll discuss next, but back to the oversize stud. How much oversize depends on the stud size you'll be using. There are three available. The standard stud has 0.3717–0.3721-in. diameter measured 1-1/2-in. up from the bottom. Oversize studs are available in 0.006-, 0.010- and 0.015-in. over the standard diameter. The best policy to follow when choosing which oversize to use is use the smallest one possible. This means don't go to 0.015-in. oversize when a 0.006-in. oversize stud will do the job.

To install an oversize stud, the hole in the cylinder head must first be reamed to size. The finish diameter should provide for approximately a 0.0007-in. interference fit between the stud and the hole. When reaming the hole, don't start with the finish-size reamer. Gradually work up to the finished size. This means you'll either have to use more than one reamer or a special step-type reamer made especially for this application. Consequently, installing an oversize stud can be an expensive proposition if you don't have access to a tool room or a kit which your Ford dealer can get especially for this purpose—which I might add will cost $30 compared to the approximate $1.50

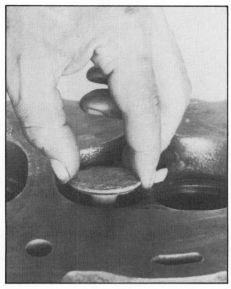 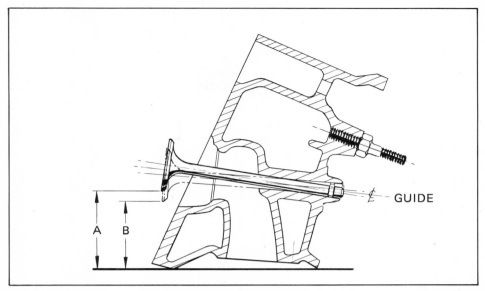

You can feel if a valve guide is worn by just wiggling a valve in the guide. To determine just how much one is worn, install a valve in the guide so its tip is even with the top of the guide. Measure how much it wiggles—the difference between A and B. Divide this figure by 3.5 you'll have the approximate stem-to-guide clearance. Measure the wiggle sideways too.

cost for an oversize stud. Once you have the hole reamed to size, follow the same procedure as outlined for installing standard studs.

Screw-In Studs—Now, for the method I prefer—even for replacing a stud which doesn't require reaming. That is to tap the stud hole using a 7/16-14 tap and install a screw-in stud. There is a danger with using this method. The tap **must** go in straight otherwise the stud will lean, causing the rocker arm to cut into the side of the stud.

Two basic types of screw-in studs are available. The aftermarket straight-shank type has to be installed by jamming two nuts together on the rocker-arm end so there will be something to thread the stud in the hole with and tighten it down. The HP289 stud, PN C30Z-6A527-B, is more convenient to install because it has a built-in nut. This, however creates a problem because the rocker-arm stud boss must be milled 1/4 in. lower so the rocker arm can be correctly positioned. This'll require the services of a machine shop.

Both of these screw-in studs are compatible with the early-style rocker arms and the rail-type rockers. You'll have to use the larger 3/8-24 lock-nut if you are replacing a positive-stop type stud and adjust the one valve using the nut. You'll end up with a positive installation, avoiding the hassle of reaming and pressing in the new stud.

Seal the Threads—Screw-in rocker-arm studs go through into the water jacket. You must use sealer on the stud threads or coolant will leak out of the head, ending up in your oil. Ford studs are hard to screw in because the threads are tight fitting so they'll lock in the head auto- matically. Aftermarket manufactured screw-in studs fit loosely, so it's a good idea to use a locking sealer such as Loctite Lock N'Seal® compound. Torque the stud to 60—70 lbs.

VALVE GUIDES AND VALVE STEMS

Valve-guide inspection and machining marks the beginning of hard-core cylinder-head machine-shop-type work which carries on over into valve and valve-seat reconditioning. This work requires special equipment and special skills. If done incorrectly, it will ruin what would otherwise be a successful rebuild. If a valve guide is excessively worn, it won't *guide* the valve so it can close squarely on its seat. Consequently, a valve can't seal properly as it will wiggle and bounce on its seat before closing. This eventually *beats out* a valve seat and accelerates guide wear. Secondly, a worn guide lets too much oil pass between the valve stem and guide, resulting in excessive oil consumption.

With this in mind, I'll cover the basics involved in reconditioning your heads, valve guides, and seats and valves.

Measuring Valve-Guide Wear—There are four ways to measure valve-guide wear: with a dial indicator at the tip of a valve which is in place in the guide, by *wiggling* a valve in the guide and measuring its movement, with a *taper pilot* and a set of micrometers or with a *small-hole gauge* and some micrometers.

To check guide wear or clearance with a dial indicator: Install a valve in the guide. Mount the dial indicator 90° to the valve stem in the direction you want to measure wear. Maximum wear usually occurs in the plane of the valve stem and the rocker-arm pivot. It can be in the opposite direction, or in the plane running lengthwise with the head if rail rocker arms are used. With the valve about 1/8 in. off its seat and the indicator tip—it should be a flat one—contacting the valve stem close to the top of the guide, hold the stem away from the indicator and zero the indicator dial. Push the tip end of the valve stem toward the indicator and read the clearance directly.

Wiggling a valve in its guide to determine stem-to-guide clearance requires a minimum amount of equipment. It's not the most accurate method, but it's good enough to determine whether your guides need further attention. It's done by measuring how much a valve wiggles, or moves at its head when pulled out of its guide so its tip is flush with the top of the guide. The amount of wiggle divided by 3.5 is *approximately* the stem-to-guide clearance in the direction of wiggle, or movement. Measure valve *wiggle* with a dial indicator or the depth-gauge end of a vernier caliper.

Using a taper pilot is the least accurate way of determining stem to guide clearance. A taper pilot is a tapered pin which is inserted into the end of a guide until it is snug. The diameter where the pilot stops is measured to determine guide diameter at the top or bottom, and compared to what its diameter should be to determine wear. This sounds great except for one thing. The pilot measures the *minimum distance* across the guide rather than the *maximum distance*. It's the maximum distance you should be looking for, so don't use this method.

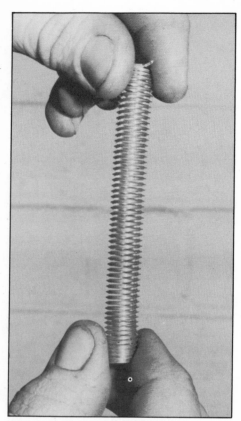

A screw-in bronze valve-guide insert. The guide being inserted must be tapped so the guide can be threaded in. Once the insert is in place it is trimmed, locked in place and reamed to size. Photo by David Vizard.

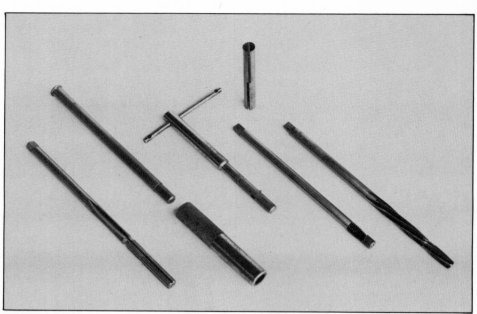

Thin-wall bronze valve-guide insert and the tools required to install it.

The most accurate method of determining stem-to-guide clearance, or simply guide wear is to use a small-hole gauge. You'll want the 0.200—0.300-in. range— the C-gauge. Unfortunately, a small-hole-gauge set costs about $30, and that's the only way you can buy them, in a set. However, if you have access to one or have decided to make the investment, here's how to use it.

Insert the ball-end of the gauge in the guide and expand it until it fits the guide with a *light drag*. Check the guide bore at several depths and in several planes to determine maximum guide wear. After you have the gauge set, withdraw it and mike the ball end. This will give you a *direct measurement* of the guide-bore cross section at the specific point. Subtract the stem diameter of the valve to be installed in the guide and you'll have the exact stem-to-guide clearance at the point of maximum wear. Measure around the guide at its top and bottom to determine its maximum and minimum diameters. Subtract the two numbers. If the difference, or *out-of-round* exceeds 0.002 in., your guides should be reconditioned.

How Much Guide Clearance?—The upper clearance for valve guides and stems is 0.005 in. The range for new stem-and-guide clearance is 0.0010—0.0020 in. for the intake and 0.0015—0.0025 in. for the exhaust. The limit you set for guide wear before redoing your guide should not be the maximum 0.005-in. limit, but somewhere between this figure and the average-standard stem-to-guide clearance—approximately 0.002 in. for both intake and exhaust valves. The number you decide on should be determined by the number of good miles you expect from your engine after the rebuild. By comparing the actual valve-guide clearance to the average standard clearance and the maximum wear limit, you can judge how many miles are left on your guides before the limit is exceeded. When doing this keep one very important point in mind. As a valve stem and guide wear, they do so at an *increasing rate*. In other words, the more they are worn, the faster they will wear. A guide and stem that has 0.003-in. clearance may very well wear at twice the rate as one having 0.002-in. clearance. Consider the valve seats too. A valve that is sloppy in its guide will wobble when closing, resulting in the seat getting beat out. Incidently, valve-guide wear is higher in an engine used for stop-and-go driving, so all things considered it takes good old "horse sense" when determining when it's time to redo your guides.

Valve-Guide Reconditioning—Valve guides can be restored by several different methods ranging from one I consider to be less than satisfactory to one that results in better-than-original valve guides. In the order of *increasing* desirability: knurling, oversized valves and guide inserts.

Knurling the Guides—Knurling involves rolling a pattern into the existing valve guide which displaces material and raises a pattern in the guide. This reduces the *effective* diameter of the valve guide, making it smaller than the original guide. By effective diameter, I mean the diameter inside the guide between the top of the displaced material, or *bumps* on one side, to the top of the bumps on the other side. Because the valve guide is now effectively smaller, the guide must be reamed to standard size. After knurling, the guide will consist of a series of peaks and valleys with the peaks smoothed off to act as the guide surface.

The problem with knurling is it's the cheap way out. And as is usually the case, it's more cheap than effective. The reason is the actual contact area between the guide and valve stem is substantially reduced, causing a proportionate increase in stem-to-guide contact pressure. Consequently, the guide wears faster. Therefore, the oil consumption problem caused by worn guides will not be cured, just delayed.

Oversize Valves—A fix which will restore your guides and valves to like-new condition is to ream the guides oversize and install valves with larger-than-standard diameter stems. The obvious disadvantage is you'll have to purchase all new valves since they generally wear about the same—exhaust-valve guides wear a little more. A set of new valves can cost up to $120—not including the cost of reaming the guides or the cost of a reamer to do it yourself. However, if some or most of your valves are not reusable for one reason or another, this is a good way to go. Ford, as well as replacement parts manufacturers such as TRW, supply valves with oversize stems.

Valve-Guide Inserts—Valve-guide inserts can restore the guides to as-good-as or better-than original condition, depending on the type used. Two basic materials are used for guide inserts, cast iron and bronze. Cast-iron inserts restore the guides to original condition. They are installed by driving them into place after machining the original guide to a diameter equal to the OD of the insert, less a couple thousandths of an inch interference fit to lock the guide into place. After the guide is installed it may or may not have to be reamed to size, depending on the way the insert is manufactured. Similar to cast-iron guides, bronze-silicone guides are much more expensive as well as more durable—normally good for well over 150,000 miles. They don't need valve-stem seals and are relatively easy on valve stems as well. Bronze-silicone guides install in the same manner as cast-iron guides.

As for bronze inserts, they will give better than original service. They come in two styles, the *thread-in* type and the *press-in* type. The thread-in type is similar to a Heli-Coil® replaceable thread except when used for valve-guide applications, it's made out of bronze with a thread form only on its OD and not the ID because it's going to function as a guide rather than a thread. To install the thread-type, a thread is tapped into the existing guide bore. A sharp tap must be used for doing this or the guide insert will not seat completely after being installed, and the insert will increasingly move up and down with the valve and actually pump oil into the combustion chamber. Consequently this type of insert requires particular care when installing. After tapping, the insert is threaded into the guide and expanded to lock it into place so the valve and the insert can't work sideways in the thread and so the backside of the thread will be in total contact with the cast-iron material of the cylinder head for maximum valve cooling. The guide is reamed to complete the installation.

The K-Line press-in bronze valve-guide insert is installed in a similar manner to the cast-iron type of insert. The existing guide is reamed oversized and the thin-wall—approximately 0.060-in. thick—is installed with a special driver. Once in place the excess insert material is trimmed off. The guide is expanded and reamed to size using the *original* valve seat as a pilot to prevent *tilting* the valve. This should be done when reaming any type of guide.

VALVE INSPECTION AND RECONDITIONING

After getting your valve guides in shape, next on the list are the valves themselves. You should have already checked them for obvious damage such as burnt heads and excessively worn tips, particularly if your engine has rail-type rocker arms—and even more so if you have a '66-1/2 to '67 289 or a '68 to '68-1/2 302. Remember, they didn't use the hardened-steel caps on the exhaust-valve tips so their wear rate was higher. Consequently, you may have eliminated some valves from the reusable list by now. The final check is for valve-stem-diameter wear. Like valve guides, valve stems also wear, but at a lower rate.

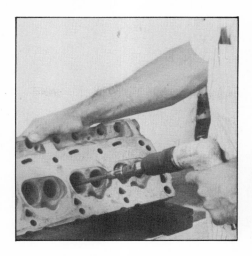
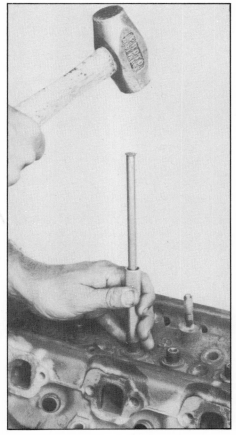
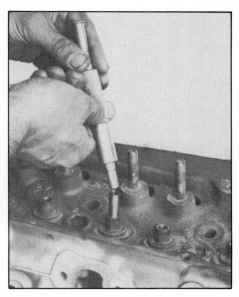

When installing thin-wall-type valve-guide inserts, the existing guides must be reamed oversize to accept the inserts. An insert is placed on the installing mandrel to prevent collapsing during the installation. It is then lubricated with anti-seize compound and driven in until it is flush with the bottom of the guide. The mandrel is removed and the excess trimmed off even with the top of the guide boss.

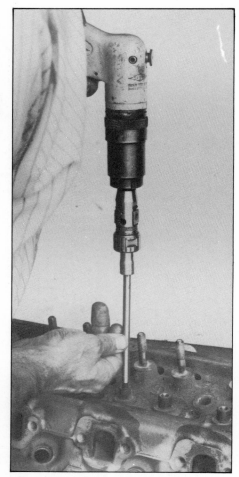 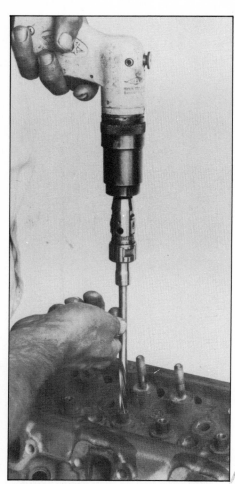

After a thin-wall valve-guide insert is in place and trimmed, it is locked into place with an expansion tool. Then the job is completed by reaming the guide to size. The result is a better-than-new valve guide.

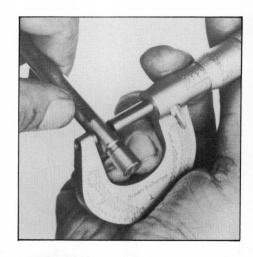 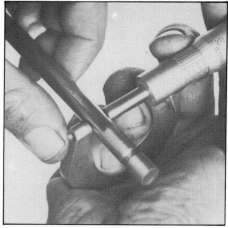

Measuring valve-stem wear with a 1-in. micrometer. Compare readings from worn and unworn portions of the valve stem. 0.0002-in. wear is way below the 0.002-in. limit, so this valve can be reused after reconditioning.

Measure Valve-Stem Wear—221 and 260 engines for 1962–1963 model years used 5/16-in. valve stems, or more precisely, minimum diameters of 0.310 in. for the intakes and 0.309 in. for the exhausts. All small blocks beginning in 1964 went to the larger 11/32-in. nominal stem diameter. The minimums are 0.342 in. on intakes and 0.3415 in. on exhausts. Use these figures to check for wear, or you can compare the *worn* and unworn portion of each valve stem to determine exact wear. Maximum valve-stem wear usually occurs at the tip end of a valve's guide-to-stem contact surface. This area is easily recognizable as the shiny portion of the valve stem. There's a sharp division between it and the unworn surface. Maximum valve opening is represented by the end of the shiny surface at the tip end of a valve. This is where a valve stops in its guide at full-open position.

To check valve-stem wear, you'll need a 1-in. micrometer. To determine stem wear, measure the unworn diameter immediately below this line. Subtract one figure from the other to determine stem wear.

Now that you have measured valve-stem wear, the question is how much wear is acceptable? Again, this depends on the service you expect from your engine after the rebuild, if and how you reconditioned the valve guides and a myriad of other questions that makes arriving at an exact wear figure too complex—and impossible. However, concentrating on desired service and the guides, should provide a satisfactory answer.

First, taking both ends of the spectrum, if valves with more than 0.002-in. stem wear are installed in guides with more than the 0.005-in. stem clearance limit or in guides reconditioned by knurling, the time spent on head work to this point could've been saved by leaving the head alone. Knurling is a Band-Aid® approach instead of a true rebuilding method. It's like wrapping tape around a radiator hose when it fails, instead of replacing the hose. The problem is not cured, it's just delayed. On the other hand, new valves or used ones with no more than 0.001-inch stem wear installed in guides reconditioned with bronze inserts should provide better-than-original service. I suggest you install guide inserts if your guides are worn excessively, preferably thin-wall bronze. Install valves with *no more* than 0.002-in. stem wear, and can be reconditioned at their tip ends without removing too much

 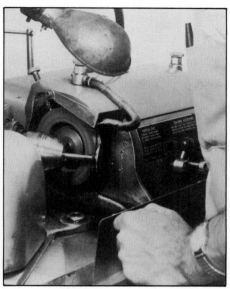

A valve being reconditioned. Tip is reground square to its stem—just enough to remove any signs of wear. Face is ground to a 44° angle to create a positive seal with the 45° valve seat.

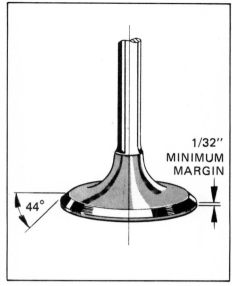

If, after grinding a valve, you find it has less than a 1/32-in. (0.031 in.) margin, the valve should be replaced. Otherwise you run the risk of burning a valve.

material. Set this as your minimum when deciding on which avenue to take when dealing with valve-stem and valve-guide wear. It should give *at least half* the durability of new valves and guides.

Reconditioning Your Valves—Assuming the majority of your valves checked out OK, the next step is to have them reconditioned. Grind their *faces* and tips. The face is the valve surface which contacts the valve seat in the cylinder head. A perfect seal must be made so the combustion chamber is sealed from the intake and exhaust ports when the valve is closed by the spring.

Valve faces are ground on a special grinder made just for this purpose. The valve is rotated in a *collet-type* chuck at an angle of 44° to a fast-turning grinding wheel. As the valve face rotates against the wheel's rotation, it is also oscillated across the face of the grinding wheel. Simultaneously the valve and stone are bathed in cutting oil for cooling and to wash away the grindings. Enough face material is removed *only* to clean the valve face. If too much material is removed the valve's *margin* will be thinned excessively and the valve will have to be discarded. The margin of a valve is the thickness at its OD, at the outer edge of the face.

A valve with little or no margin will approach being sharp at the outer edge of its face and must be replaced because it's highly susceptible to being burned, particularly if it's an exhaust valve. An exhaust valve should have at least a 0.030-in. margin, or approximately 1/32 in. if you like fractions. Because an intake valve doesn't run as hot as an exhaust valve, its margin can be as narrow as 0.015 in.—approximately 1/64-in.

After the valve face is ground, the tip is faced off, or ground squarely to the centerline of the valve using another attachment on the valve-grinding machine. Next, the tip edge is beveled, or *chamfered* to eliminate the sharp edge developed during the previous operation. Your valves should now be as good as new and ready for assembly into the cylinder heads—when they are ready. It's now unnecessary to keep your valves in order because all traces of mating to other components are gone.

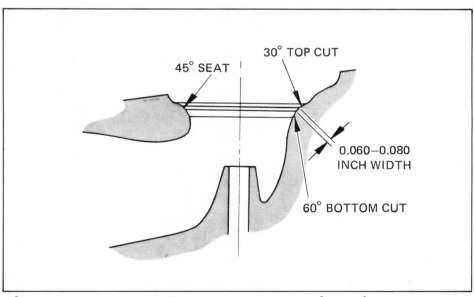

45° valve-seat inside and outside diameters are established by 30° and 60° top and bottom cuts, respectively. A seat's outside diameter should be approximately 1/16-in. smaller than the outside diameter of the valve's face. Top and bottom cuts also establish seat width.

 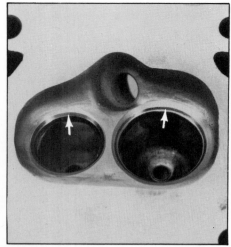 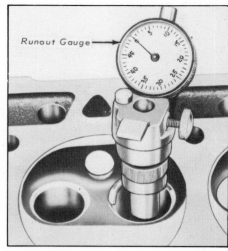

Grinding valve seats and the tools needed to do it: Three well dressed grinding stones with angles of 30°, 45° and 60°, a pair of dividers for measuring seat diameter, a ruler for setting the dividers and measuring seat width and some blue dye so the seat will show up. Dark portions (arrows) are the 45° valve seats.

After a valve seat is ground, its concentricity or runout in relationship to the valve guide should be checked with a gauge such as this. Photo courtesy Ford.

VALVE-SEAT RECONDITIONING

Valve seats are reconditioned just like the valves, by grinding. Equipment used for doing this is special, expensive and requires some degree of skill and experience, so valve-seat grinding is a job for the specialist. Also, it's a job which is a necessary part of every rebuild. This is particularly true for heads having reconditioned valve guides because the valve seats will not be *concentric* with the new guides—they won't have the same centers. Consequently, even though a valve seat may be in good condition, the guide won't let the valve seat evenly when closed. It will be held off-center in the seat and cannot seal, nor can it fully close. Because a valve seat is ground using a mandrel located in the valve guide, seat concentricity will be corrected.

Grinding the Seats—Valve seats are ground using manual, electric- or air-powered grinders alternating between three grinding stones with angles of 30°, 45° and 60°. Power grinders are best because they provide the most accurate seat. The most important angle, 45°, is the actual valve-seat angle. If this 45° figure strikes you wrong, it's because the valves were ground to a 44° angle rather than 45°. This one-degree difference provides an *edge-contact* at the outer periphery of the valve face so hot combustion gases will be sealed the first time the valve closes. Leakage at either valve due to poor seating can warp the valve, and burn the valve face and seat.

The other two, 30° and 60°, are the *top cut* and the *bottom cut*, respectively. The top cut is on the combustion-chamber side of the valve seat. It trues up and establishes the outside diameter of the valve seat. Valve-seat OD should be ground approximately 1/16-in. smaller diameter than the valve OD. The bottom cut 60° angle on the port side of the valve seat establishes the seat ID, but more importantly, the valve-seat width. After the top cut is made, the valve seat is narrowed by taking more bottom cut.

Seat widths of 0.060–0.080 in. are used for all small-block valves. For cooling reasons, exhaust-valve seats are generally ground to the high side of the range (wider) to provide additional valve-to-seat contact to ensure valve-to-seat heat transfer and to reduce "beating out."

Tools which should be on hand to assist in the grinding process, other than the actual grinding equipment, are a set

Lapping valves using lapping compound between the valve face and seat. Valve is inserted in its guide, the lapping tool is attached to the head of the valve, then the valve is rotated back-and-forth while being held lightly against its seat. Consider this operation unnecessary because it won't correct an incorrectly ground valve or seat and it's not necessary if they are done right. It will, however show up a bad valve job real quick.

of dividers, Dykem-type metal bluing, a scale for setting the dividers for measuring seat OD and directly measuring valve-seat width and a special dial indicator for checking valve-seat concentricity. The bluing assists in distinguishing the valve-seat edges from the top and bottom cuts. It is applied to the seat prior to making these cuts. Consequently, the seat shows up as a blue ring contrasted against the brightly finished top and bottom cuts making the seat easier to see and measure.

Hand-Lapping Valves—*Lapping valves in* is nothing more than grinding a valve face

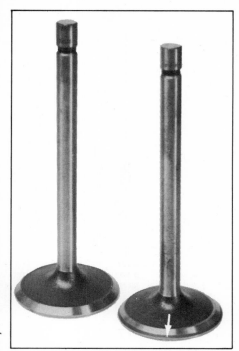

Gray band (arrow) on the lapped valve shows the valve-face-to-seat contact area.

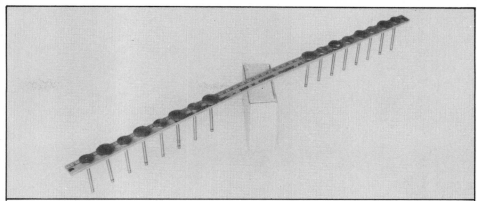

Valves should be installed on the seat they were lapped in on. Keep them in order as I'm doing here with a yard stick.

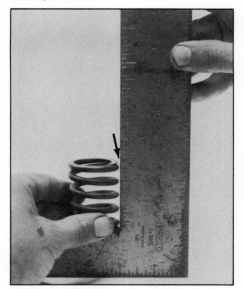

Checking a valve spring for squareness with a carpenter's framing square. It should lean no more than 1/16 in. at the top (arrow).

Checking spring load at installed height. Shimming is done to bring spring load within specification, then the spring is checked at open height to ensure it doesn't go solid. You should be able to see daylight between all the coils at open height.

and its seat *together. Lapping paste* is applied to the valve face, then the valve is rotated back-and-forth by hand in a circular motion while simultaneously applying a light pressure to the head of the valve.

I'm mentioning valve lapping in case you are aware of the process and think it is necessary for valve sealing. Generally speaking, if a valve face and its seat aren't ground correctly, lapping won't fix the problem, and if they are done right, lapping isn't necessary for proper valve seating and sealing. So don't waste your time with lapping. A rebuilder with a good reputation for engine work that performs and lasts is your best guarantee.

Do They Seal?—It would be nice to know if your valves are going to seal before installing the heads on your engine. This is so you can do something about it now if there is a problem. To determine if your valves are sealing OK you'll first have to assemble the valves in the heads. Either do it temporarily now, then disassemble the heads for checking and setting up the valve springs, or complete the valve-spring checking and setting-up as outlined in the next few pages, then assemble the valves in the heads and check them.

With the valves in place, position the heads upside down on your bench so the head-gasket surface is level. Using kerosene, fill each combustion chamber and check inside the ports for leaks. If you don't see any leaks you can consider the valves OK, otherwise you'll have to confer with your rebuilder to determine what the sealing problem is, then have it corrected.

VALVE SPRINGS— INSPECTION AND INSTALLATION

After your heads and valves have been completely reconditioned, the valve springs must be inspected. A valve-spring tester helps, but it's not absolutely necessary. It's helpful to understand what's involved in valve-spring testing. You should become familiar with valve-spring terms and the importance of maintaining certain standards. The terms are; *spring rate, free height, load at installed height, load at open height* and *solid height*.

Spring Rate—Spring rate is not one of the commonly listed valve-spring specifications as such, but relates directly to most of the other specifications. Also, it's one of the basic terms necessary to describe a coil spring's mechanical properties.

Spring rate governs the load exerted by a spring when compressed a given amount, usually expressed as so many pounds per inch of deflection. When a spring is compressed some amount, it requires a given amount of force to do so. Typical valve-spring rates vary between 200 and 400 pounds per inch. Spring rates are usually indicative of the RPM range of the engine they are installed in. For example, the conventional small-block Ford valve-spring rate is approximately 250 lb./in., whereas the higher-revving HP289 spring rate exceeds 350 lb./in. to control the additional valve-train inertia. The higher spring capacity is required just to close the valves, or keep them from *floating* when the engine is operated in the high RPM ranges. If a valve spring loses its rate, or *resiliency*, it is no longer capable of closing its valve at or near maximum rated RPM.

Free Height—Free height or length of a valve spring is its unloaded height or length. If a spring's free height is too long or too short, the load exerted by the spring as installed in an engine will be off because it will be compressed more or less by the spring retainer. Therefore, knowing what spring rate is in addition to free height gives you a clue as to why a spring does or doesn't meet its specifications.

Load at Installed Height—A spring's load at its *installed height* is a common spring specification—installed height is spring height as installed in the cylinder head; measured with the valve closed. This is the distance from the cylinder-head spring seat to the underside of the spring retainer. A typical load-at-installed height specification is 70–80 pounds @ 1.750 in. When compressed to a height of 1.750 in. from its free height, the load required to compress this spring when it is new should be between 70 and 80 lbs. The *absolute minimum limit* is 10% less than the *minimum standard limit*, or 61 lbs. in this case. If a spring does not exceed or at least meet the minimum it should be replaced. If your engine will be operated at its upper RPM limit or is an HP 289 or Boss 302, you should restrict valve-spring load to the *minimum* standard limit.

Load at Open Height—Another commonly listed specification is a spring's compressed load when its valve is fully open. A typical specification is 160–175 lbs. @ 1.400 in. Just as with the installed-height load, a spring must fall within the 10% minimum-load limit or be replaced. In this case the minimum is 144 lbs. Again, consider how your engine will be operated when checking the springs.

Solid Height—Solid height describes the height of a coil spring when it is totally compressed to the point where each coil actually touches the adjacent coil. Coil springs should never be compressed to this height in normal service. If a valve spring were to be compressed to its solid height prior to the valve opening fully, the load on the valve train would theoretically approach infinity. But before this could happen the weakest component in the valve train will fail—something bends or breaks. The usual result is a bent pushrod.

Squareness—*Valve-spring squareness* is how straight a spring stands on a flat surface, or how much it *tilts*. It is desirable for a valve spring to be square so it loads the spring retainer evenly around its full circumference. Uneven retainer loading creates additional valve-stem and guide side loading with a corresponding increase in stem and guide wear—something you should minimize. Limit out-of-squareness to 1/16 in. measured at the top of the spring to a vertical surface.

Checking the Springs—Of the spring characteristics just covered the ones which should be checked are squareness and the installed and open spring loads, plus making sure the spring doesn't

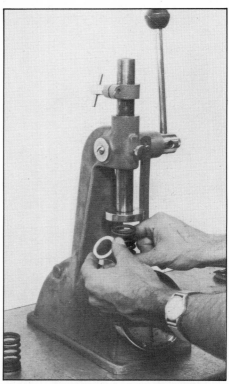
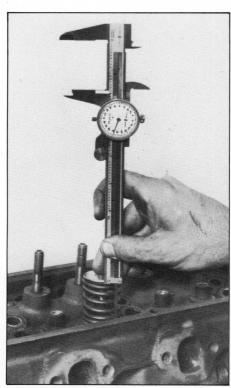

Checking installed spring height with the depth-gauge end of a vernier caliper. Measure from the spring pad on the head to the underside of the spring retainer. Add spring shims to make up the difference.

reach its solid position in the full-open valve position. A flat surface, square and something to measure with is all that's needed to check spring squareness. When you're all set up, rotate the spring you're checking against the square to determine its maximum tilt, then measure it. If the gap at the top of the spring exceeds 1/16-in., replace it. As for the last two checks, a valve-spring tester is required. Two types of testers are normally used, but both do the same thing. They permit manual compression of a spring to the opened and installed heights so the spring's loads at these heights can be read. The question is, "How do I check for a spring's loads if I don't have a spring tester?" The answer is simple, you don't!

Fortunately, a simple method can be used to determine if a spring may be faulty. That is to compare its free height to that originally specified. The reason this works is if a spring has been fatigued or was overheated, usually from excessive engine temperatures, the spring will *sag*, or collapse some amount. This will be reflected in a spring's free height. The result of a sagged spring will be a reduction in spring load in the installed and open positions. Also, when a spring is removed from its cylinder head, it will not return to its original, or specified free height. Use the valve-spring chart to determine what the free height of your springs should be, then measure them. Generally a valve spring should be within

0.100 inch of the specified height. Otherwise they should be checked with a spring tester to make sure they will provide adequate seated and open pressures.

Valve Springs—Replace or Not to Replace?—Generally, I replace any spring which is 1/8-in. shorter than the specified free height, regardless of the planned engine service. A valve spring that's 1/16-in. shorter can be installed in an engine that's to be used for light service. This decision should be based on what type of service your engine will be used for and how far off your springs are. So, if it's going to be used for puttering around town, you probably can get away with marginal springs. This is not the case though if the engine is to be operated at or near its peak RPM. This being the case, the springs will get progressively weaker and those valves with weak springs will float at a lower engine RPM than the engine may be able to turn otherwise.

If your engine is to be used for severe service, the springs should be tested on a spring tester to determine if they can be corrected by inserting a *shim* under them if any have a *short* free height. In this use, the shim—a special flat washer—is placed between the spring and the cylinder head so the spring will be compressed more to restore the spring's installed and open loads. However, if not used carefully, shims can cause severe valve-train damage by making the spring reach its

VALVE SPRING SPECIFICATIONS

Engine	Year	INSTALLED Height (in.)	INSTALLED Load (lbs.)	INSTALLED Min. Load (lbs.)	OPEN Height (in.)	OPEN Load (lbs.)	OPEN Min. Load (lbs.)	Free Length (in.)	Solid Height (in.)	Color Code
221, 260	62-63	1.77	57-63	51	1.38	161-178	145	2.15		Lt. Brown
260, 289	63-66½	1.78	71-79	65	1.39	160-178	145	2.09	1.30	Blue or Grey
289	66½-67½	1.64	57-63	52	1.25	158-174	143	1.86	1.11	Black or Grey
	67½-68	1.66	71-79	64	1.27	174-192	157	1.93	1.21	Black or Grey
HP289	#63-67	1.77	83-89	75	1.32	234-260	211	2.02		Red
302	68-69½	1.66	71-79	64	1.27	174-193	157	1.93	1.21	Black or Grey
	69½-72	1.69	76-84	69	1.31	190-210	171	1.94	1.23	Lt. Red
	73-75½	1.69/1.60	76-84	69	1.31/1.22	190-210	171	1.94/1.85	1.23/1.16	Lt. Red/Purple
	75½-78	1.69/1.60	76-84	69	1.31/1.20	190-210	171	1.94/1.87	1.23/1.14	Lt. Red/Yellow
Boss 302	#69-70	1.82	88-96	80	1.32	299-331	270	2.03	1.27	4 Blue Stripes
351W	#69-70½	1.79	79-87	71	1.34	204-226	184	2.07	1.27	Lt. Blue
	70½-76	1.79	71-79	65	1.34	190-210	171	2.06	1.24	2 Green Stripes
	77-78	1.79/1.78	71-79/74-82	65/67	1.34/1.36	190-210 / 194-214	171/175	2.06/2.04	1.24/1.29	2 Green Stripes/Yellow

If specifications are different for intake- and exhaust-valve springs, intake-spring will be shown first (intake/exhaust).
Used with a damper spring.

Use this chart for reference when checking your springs.

solid height before its valve is fully opened. Also, another problem when using valve-spring shims is a spring is designed to be compressed a certain amount *continuously*. This amount is the difference between its free height and open height. If it's compressed more than this amount the spring will be overstressed, or overloaded resulting in an over-worked, or fatigued spring. The spring will quickly lose its load-producing capacity. So if you have a questionable spring, replace it rather than shimming it, even though a new spring costs five times more than a shim. If you do shim some of your springs, make sure you attach the shim to its spring with some wire, string or whatever. If they get mixed up you'll have to retrace your steps and recheck them.

ASSEMBLE YOUR CYLINDER HEADS
Now that all your cylinder-head parts have been inspected and/or reconditioned, you can get them all back together again—sort of like Humpty Dumpty. Except in this case you won't need any of the King's Horses or the King's Men, just a valve-spring compressor and a ruler, plus some valve-spring shims—maybe. Yes, you may still have to install shims even though you haven't had to up to this point.

Valve-Spring Installed Height—One of the valve-spring loads you checked was the spring's *installed height*, or the height the valve spring should be with the valve closed and the spring and its valve, retainer and shim/s installed. Shims are included in the spring's installed height *if used to correct spring free height*. Make sure every spring is at its correct installed height, otherwise its valve will not be loaded correctly.

The position of a valve seat determines the amount the valve stem will project out the top of the guide when the valve is closed. It also follows that the distance from the cylinder-head spring seat, or spring pad to the underside of the valve-spring retainer, as installed on the valve, is also determined by the valve seat.

As originally manufactured, this distance, or pad-to-retainer dimension, coincides with the valve-spring installed height. However, because material is removed from each valve face and seat during the grinding process, this distance is increased and consequently exceeds the installed spring height. If your valve seats were in good condition, only a small amount of material should've been removed to clean them up. Therefore, the spring's installed height should not have been affected to any great degree. On the other hand, if more material was removed, the spring's installed height and load could be changed considerably, so they should be checked.

Because setting and checking installed spring height is best done as part of the cylinder-head assembly process, let's get on with putting your heads together and doing the height checks concurrently.

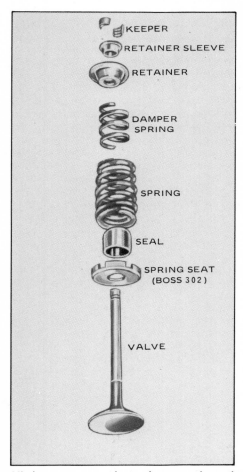

All the components that make up a valve and spring installation. Damper spring is unique to the HP289, Boss 302 and early 351W engines. Separate spring seat is only used in the Boss 302. A two-piece spring retainer is shown. Photo courtesy Ford.

89

After lubricating the valve stem, insert the valve in its guide. Hold it in the closed position while installing the seal. You can now loose-assemble the spring seat—if you have a Boss 302—any shims, the spring and retainer. With a compressor, pull the spring and retainer down far enough so you can install the keepers. Release the compressor and check to make sure the retainers are locked into position.

Using a brass punch to install the freeze plugs. Some silicone sealer around its edge will ensure an immediate seal.

Have your valves ready to be installed in the heads *in order* and the springs with their shims, if you used any to correct valve-spring free height.

Installed spring height can be checked by two methods. The first is by simply inserting the valve in the head, then assembling the retainer and keepers on the valve stem without the spring, shims or valve-stem seal. The retainer is lifted up to hold the valve in its closed position—it also keeps the retainer and keepers from falling down the stem—while the seat-to-retainer dimension is measured. To get an accurate reading, the retainer must be lifted up firmly to ensure the valve is against its seat and the retainer and keepers are in place. A *snap-gauge* and micrometers will give you the most accurate reading when measuring this dimension, however you can do pretty well using a six-inch scale if you have sharp eyes. One trick you can use is to cut, then file a short section of welding rod or *heavy* wire to the specified installed height of the spring. You can accurately check the rod length using a micrometer. This will give you an accurate *gauge* which when placed between the spring seat and the spring retainer will show a gap the same thickness of the shim if one is required. The shim that fits between the end of the rod and the retainer or the spring seat is the one to use.

If you've used the measuring method to determine shim thickness, after you've arrived at a shim-pad-to-retainer figure, subtract the installed spring height from it and you have the shim thickness required to provide the correct installed height. Therefore:

Shim thickness
= spring-to-retainer dimension
− installed spring height

Shims come in thicknesses of 0.015, 0.030 and 0.060 in. They can be stacked in varying combinations so you'll end up with the right overall thickness.

The second method of checking seat-to-retainer distance varies only in that you install the spring too. This means you'll have to use the spring compressor to do so, but it does ensure the reading will be accurate. The valve will be closed and the retainer and keepers will be fully seated. If this method is used you won't be able to use a *snap gauge*. You won't be able to get under the spring retainer because the spring is there. One measuring device that does work well here is a pair of *vernier-type calipers*. You can also use your less-accurate six-inch scale. Measure the actual spring height in this case by measuring from the spring pad to the top of the spring. Regardless of the method used, the way of arriving at shim thickness is the same—subtract *specified* installed spring height from the dimension you just measured. If the variation is less than 0.015 in., you don't need a shim.

When checking installed spring heights, keep a record of the shims and their location as you go along. Also, line up the valves with their respective shims, springs, retainers, keepers and seals so they'll be organized for assembly time. Remember, you can mix up seals, retainers and keepers, but the *springs, shims and valves must be installed in the position in which they were checked.*

After finishing the installed-height check use your list to purchase the necessary shims. While you're at it, pick up four freeze plugs if you haven't done so already. You'll need the 1-1/2-in. diameter cup-type plug—Ford D7AZ-6026-A. You should now have all the parts necessary to assemble your cylinder heads.

Before assembling your heads make sure all the parts are clean. Using your shim list, organize the shims you just purchased so they'll be installed in the right location when it comes time to install them. One thing that's handy to know when looking for a particular thickness shim is they are color-coded according to their thickness. For example, V.S.I. uses the following colors: Black, 0.015 in.; silver, 0.030 in.; gold, 0.060 in

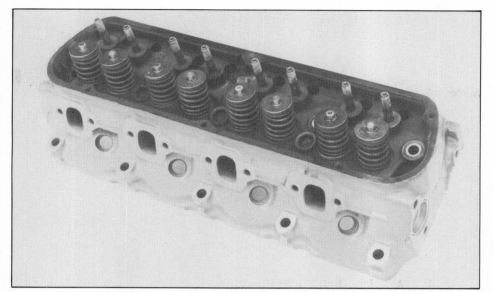

A completely assembled cylinder head, and with paint too! It'll remain covered while waiting to be assembled to the engine. This is a 4-V 302 head incorporating adjustable press-in studs. Each Boss 302 head assembly includes four additional parts—pushrod guide plates which are retained under screw-in studs.

All that's involved in reconditioning an intake manifold is a good cleaning beginning with scraping off all the old gasket material.

Put the Freeze Plugs In—Start the cylinder-head assembly by installing the freeze plugs. Stand the heads on their ends to do it. Freeze plugs can be installed without sealer, but it's not a bad idea to put a small bead of silicone-type sealer around the outside edge of the hole. Sealer won't hurt anything, but how you install the plug can. Set the plug squarely over the hole so you're looking into the plug—the concave side is pointing out. Use a hammer and the largest diameter punch you have that's not over 1-3/8-in. to drive the plugs in. A punch that fits the plug ID is perfect, or a 1-in. diameter brass punch also works well—it does for me. Don't use a punch with less than a 1/2-in. diameter. It will distort the plug, causing it to leak. Likewise, don't drive the plugs in by hitting them on the edges of their flanges. The same thing will happen. Install freeze plugs by driving them in a little at a time. Work around the inside edge of the plug, making sure it goes in squarely. When the outer edge of the plug lines up with the ID of the hole, the plug is in far enough. Turn the head end-for-end and install the other plug—then do the other head.

Install the Springs and Valves—Set the heads on their sides on your work bench and start at one end of each of the heads installing its valves and springs. Limit confusion by working in the same order you used when checking installed spring heights. Oil each guide and valve stem, insert the valve in its guide and slide a stem seal all the way down over the valve's stem. This will keep the valve from falling out of its guide while you're installing the shim/s, spring, damper spring, if your engine is so equipped and spring retainer in that order. The stem seals are included in your engine's gasket set. It's possible you'll have shims for correcting spring *free height* as well as for correcting *installed height*—or you may have none at all. At this point use the spring compressor to pull the retainer down around the valve-stem tip for installing the keepers—and the retainer sleeve if your engine uses the two-piece-type retainers. Don't compress the springs more than necessary. With the keepers in place, release the spring compressor and check the keepers to make sure they're in place. Correct them if they are not and continue installing the valves and springs. If you made a gauge from welding rod or wire you can use it now to confirm installed spring height. With the installation of the last spring and valve, the cylinder heads are assembled to the point that they can be installed on the block when it's ready. Spray your heads with oil and store them covered up so they'll remain clean, dry and rust-free while you're preparing your engine block.

INTAKE MANIFOLD

All that's required to restore your intake manifold to like-new condition is a thorough cleaning. If it was hot-tanked, this should be a relatively easy job. Scrape its gasket surfaces—head, cylinder block and carburetor.

Remove the Baffle—An intake manifold has one difficult area to clean—its underside, directly under the carburetor-heat passage. This passage is built into the manifold and connects the exhaust ports for cylinders 2 and 7 for the sole purpose of providing heat for the fuel charge for cold engine operation. To clean the sludge build-up from this area you'll have to remove the baffle which shields the engine oil from the hot manifold underside. Two spiral-grooved rivets secure the baffle to

Intake-manifold heat passage connecting the number-2 and 7 cylinder exhaust ports. The baffle (arrow) shields engine oil from the manifold's hot underside. Photo courtesy Ford.

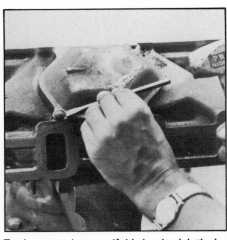  

To do a complete manifold cleaning job the heat baffle will have to be removed. This'll require a very skinny wedge to remove the rivets. Be careful when removing them. Their heads are easily broken off. This is the buildup you'll normally find under a baffle.

 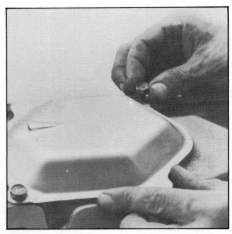 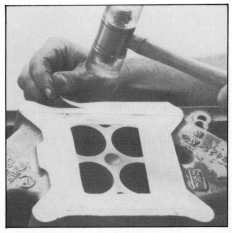

Here's the easiest way to remedy a broken rivet. Drill it out, then tap the remaining hole with a 1/4-20 tap. Install the baffle with a 1/4-20 bolt and lock washer and you're back in business. Under no circumstances should you leave the baffle off—you'll be wondering why your newly built engine is consuming oil. It'll be vaporized and burned or released into the atmosphere through your crankcase ventilation system.

Finishing off this manifold by masking off the machined surfaces with masking tape prior to painting. If you've wondered how to trim masking tape the easy way, this is how—with a soft hammer. Don't apply this method when masking sheet metal.

the manifold. To remove them you'll need to wedge something under the rivet heads, something with a very sharp wedge. A long skinny chisel will do the job. One thing to be aware of, the rivet-heads are easily broken off. If this happens you'll have to drill the remainder of the rivet out.

Wedge under the rivet-head until the head is about a quarter-inch off the baffle, or far enough out to clamp on the head with a pair of Vise-Grip® pliers. With the pliers clamped on the rivet head, pry under the pliers while turning the rivet counter-clockwise so the rivet unscrews out of the manifold. When you've removed the first rivet all you have to do to the second, and last one, is to back it out far enough so you can rotate the baffle to the side for access to remove the sludge. When you have this area well cleaned simply rotate the baffle back into place and install the rivet. Here's where skinny fingers or needle-nose pliers come in handy. Hold the rivet so it starts in squarely and lightly tap it into place.

Break a Rivet?—Because it's relatively easy to break a rivet head off don't be discouraged. Get on with the job of removing the remainder of the rivet so you can reinstall the baffle after you have the manifold cleaned. If you do break a rivet head off, don't think that one rivet is enough to hold the baffle in place or your engine can do without the baffle altogether—it can't. Rather than trying to find a replacement rivet, replace it with a 1/4"-20 x 3/8-in. long bolt or screw. First, drill an 1/8-in. pilot hole in the center of the broken-off rivet. Center-punch the rivet first to make sure the drill doesn't run off to the side. Next, use a 7/32-in. drill to remove the remainder of the rivet. A 7/32-in. diameter hole is the correct size for piloting a 1/4"-20 tap. After tapping the hole and cleaning the manifold, reinstall the baffle using bolt and lock-washer.

Engine Assembly 7

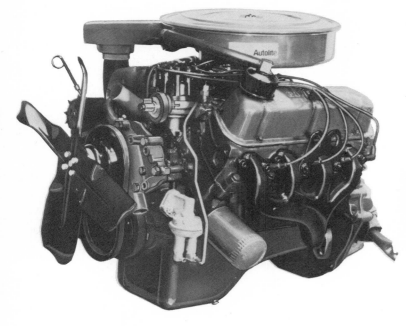

1970 351W setup for installation with a manual transmission. This engine is the big small block, used in most of the Ford car lines as well as the light trucks. Larger displacement was gained by raising the block deck-height 1.275 in. so stroke could be increased to 3-1/2-in. Photo courtesy Ford.

This is the part of the engine rebuilding process I like. Everything is clean, all the parts are new, reconditioned or have been checked out and the running around associated with getting parts and jobs done you can't handle is just about over.

Things You Need Before Starting—Just like all new jobs you've done up till now, there are certain things you'll need. One thing is all the parts. Trash bags are useful for covering up clean parts, particularly the block during assembly. You'll also need a complete engine gasket set, gasket sealer, gasket adhesive, or some weatherstrip adhesive and a spray can of aluminum paint. All sorts of sealers and adhesives are available, but I'll just list a few that work particularly well in certain applications. First is room-temperature-vulcanizing (RTV) silicone sealer. It's great if it's used right in the right place; used incorrectly it can be disasterous. You have to know its limitations.

It's not a cure-all. Ford markets some particularly good sealers. Perfect Sealing Compound, B5A-19554-A, is a general-purpose sealer. Their other one is particularly good for installing intake-manifold gaskets, Gasket and Seal Contact Adhesive, D7AZ-19B508-A. Another good one for this purpose is OMC's (Outboard Marine Corporation) Adhesive Type M. You can also use weatherstrip adhesive as a gasket-adhesive substitute. Ford's is COAZ-19552-A and 3M Corporation's is 08001.

Lubricants are a necessity when assembling an engine. How an engine's critical parts are lubricated during its first few minutes of initial run-in will be a major determining factor in the engine's durability. Remember this during the assembly process. Lubricants to have on hand include at least a quart of the oil you intend to use in your engine's crankcase—probably a multi-grade detergent oil—a can of oil additive and some molybdenum-disulfide grease. As for what brand of oil to use, I am not going to make any suggestions because the brand isn't as important as the grade. So, regardless of the brand you use, use the *SE grade*. In addition to crankcase oil, get a couple of cans of Ford's Oil Conditioner, D2AZ-19579-A, or GM's EOS (Engine Oil Supplement) for general engine assembly, initial bearing lubrication, and to put in the first crankcase fill. Finally, in the oil department, you could also use a squirt can. Fill it with crankcase oil for easy application.

Here are some adhesive-type sealers that are handy when it comes to assembling your engine.

93

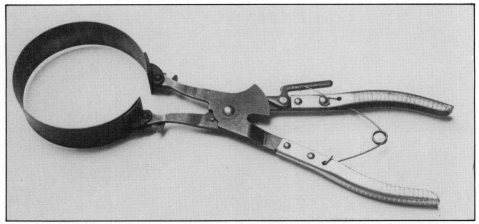

Two tools you must have when assembling an engine. Even the most experienced mechanic doesn't guess when torquing bolts. The ring compressor is a must when installing rod-and-piston assemblies.

Now, for special tools other than those normally residing in your toolbox. The first one is a tool you may have. A torque wrench. I mention it especially because *an engine cannot be assembled correctly without a torque wrench*. When it comes to tightening bolts, even the most experienced mechanic doesn't rely on feel, he uses a torque wrench. Therefore, put this tool at the top of your list. The next item is one that I don't consider necessary because bearings are made to such close tolerances, however it's not a bad idea to use it as a check to be sure you get the right bearings. It's not uncommon that the *wrong bearings* were put in the *right box*. I'm referring to *Plastigage*, a colored strip of wax used for checking bearing-to-journal clearances. All you need to know about it at this point is you'll need the green Plastigage which measures 0.001–0.003 in. Get the minimum amount when purchasing it—you won't need much. The need for the next tool depends on the route you took with your cam bearings. If you left your old ones in the block or had your engine machinist replace them for you, you won't have to concern yourself. If you have yet to replace them then you will need a cam-bearing installation tool set. Before you rush out to rustle one up I'll take this opportunity to suggest that you have an engine shop do it for you. It's not too late yet. However, if you still insist on doing it yourself, read on.

CAM BEARING INSTALLATION

So, I didn't scare you out of attempting to replace your own camshaft bearings? Well, you can't back out now so let's get on with it. Like I just said, the first thing you'll need are the tools. These come in varying degrees of sophistication, from the type which pulls the bearing into its bore with a threaded rod, nut, thrust bearing and a *mandrel* to one that drives the bearings with a mandrel, driving bar and a hammer. When installing camshaft bearings, three things must be kept in mind. First, the bearings must be installed square to their bores, the oil holes in the bearings must line up with those in the block and care must be taken during the installation so as not to damage the bearings.

They Are All Different Sizes—Even though I indicated in the teardown chapter that the five cam bearings have different bore diameters, I wasn't as explicit then as now due to the relative importance of removing the bearings as opposed to installing them. Here's a big shocker for you. All small-block Ford cam bearings are the same regardless of whether the engine is a 221 or a 351W. Their bore diameters get progressively smaller from front-to-back: 2.083, 2.068, 2.053, 2.038 and 2.023 in. Also, because the bearing shells are approximately the same thickness, their bores in the block have the same relative diameters as the bearings' outside diameters. Consequently, each bearing will fit in only one location.

Chamfer the Bearings—Before installing the bearings, it's a good idea to remove the sharp edges from the inside edges of all the new bearings. They can interfere with the cam journals as you're trying to slide the cam into place, making cam installation more difficult than it needs to be. The tool particularly suited for this is a *bearing scraper*. It's like a triangular file with no teeth. If you don't have one of these, a pen knife is a satisfactory substitute. To chamfer the bearing ID, hold your knife or scraper 45° to the bearing surface and hold the bearing so you can rotate it while peeling a small shaving about the size of four human hairs—if you can picture the size of four hairs. The idea here is you don't have to remove much, just enough so you can't feel a burr when you drag your finger nail across the edges of the bearings.

Get the Block and Bearings Ready—Position your cylinder block on its back so you'll have access for installing the bearing inserts. Also, it's best to locate the block so you can easily sight down the center of the bearing bores during the installation, particularly if you have the type of bearing installer that's not self-centering. This is to ensure the bearings enter straight.

After your block is positioned, organize the bearings in the same sequence they are to be installed in. This'll save some time fumbling around. Also, fully understand where the bearing-insert oil holes are to be positioned in the bearing bores so they'll coincide with the oil-holes. Cam-bearing lubrication depends on oil being fed up from the crankshaft main bearings. If an insert is rotated in its housing so the oil hole is closed off, the bearing and journal won't receive any lubrication. The result is a wiped out bearing and possibly a severely damaged bearing journal. Oil holes in the bearing inserts are slotted to accommodate some misalignment between them and the oil holes, but not much more than five degrees. In addition to the normal one hole for lubricating a cam journal, there are two holes in the front bearing

Each of the five camshaft bearings is a different size. They get smaller from the front to the back of the engine. Bearing numbers and their corresponding positions are printed on the side of most cam-bearing boxes. Size difference can be seen by this photo of front and rear bearings.

  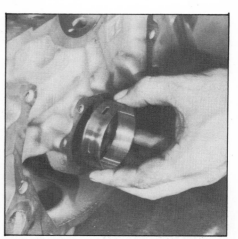

New camshaft bearings have square edges and possibly a slight burr which makes installing a cam difficult. Before installing the bearings, slightly chamfer these edges with a penknife or a bearing scraper as I'm doing here.

Holes in the cam bearings are there for a purpose, to provide a passage from the oil galleries to the bearing journals. Front bearing insert has two holes, so be careful. Extra hole is for cam-journal-to-distributor-shaft-journal oil gallery shown in the cutaway drawings on pages 52 and 72.

insert. The additional hole feeds the bottom distributor-shaft bearing. Therefore, one hole lets oil into the cam bearing and the other lets it out to lubricate the distributor bearing.

Install the Bearings—With everything ready to go, you can begin installing the cam bearings. Start with the front or rear bearing and work toward the center of the engine from the opposite end of the engine. When you finish installing the center bearing, turn around and install the remaining two bearings by also working from the opposite end. The reason for doing this is the farther you are from the bearing—relative to its position in the block—the more accurate the installation tool can be lined up with the centerline of the camshaft bearing bore. You'll have the bores at the opposite end of the block from the bearing you're installing to use as a reference to line the bearing and tool up. This is particularly true of the drive-in type which is lined up using the "eye-ball" technique.

To install a bearing, select the one which fits the location you want to start with. Although obvious, most bearing manufacturers list bearing locations with

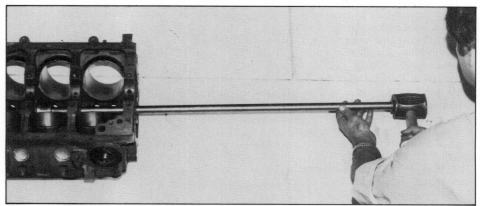

Installing cam bearings is no different than removing them, except considerable care must be taken to make sure they don't get damaged, the oil holes line up and they are square.

each part number on the box. Cross-reference these numbers with those on the bearing shells to confirm their locations if you have any doubts. Slip the insert over the mandrel. If you have the expansion-type mandrel you'll have to expand it so it fits snugly in the ID of the bearing. If you have the solid type, select the one that fits the bearing ID. With the pull- or drive-rod handy, position the bearing and mandrel over the bearing bore so the oil hole in the bearing lines up with the one in the block—so they will match after the bearing is installed. Also, if you have the threaded, or pull-type installation tool, locate the bearing and mandrel on the opposite side of the bearing web from which you'll be pulling. When using the drive-in type, you obviously install the bearing from the same side you'll be driving it in from. Regardless of the type tool you're using, check the bearing immediately after you've gotten it started in the bore to make sure it's

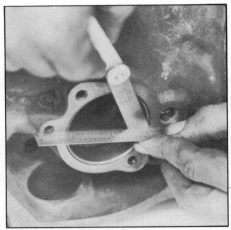

All but the front cam bearing can be eyeballed during the installation. It must be installed 0.005–0.020 in. behind the front face of its bore, being checked here with a straight edge and feeler gauge.

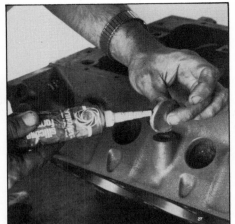

Some silicone sealer and a large-diameter punch for installation help ensure that your freeze plugs won't leak. Whatever you do, don't hammer on the edge of a freeze plug to install it. This will distort its edge and increase the likelihood of a leak.

going in straight, then finish installing it if it's OK. If not, straighten it up and finish the installation. Except for the front bearing, stop pulling or pushing the bearing in when it looks centered in its housing. As for the front bearing, it must be located more precisely.

Number-One Bearing Requires a Feeler Gauge—The front edge of the number-one cam bearing must be located *0.005-0.020 in. behind the front face of the engine block for timing-chain lubrication.* The camshaft thrust-plate-mounting surface conveniently coincides with the front-face-of-block, so you can work from it. When you think the bearing is close to its right location check it with a straight-edge laid across the thrust-plate surface. Gauge the distance between it and the front edge of the bearing. If the clearance is not within the 0.005–0.020-in. range, move the bearing accordingly. You can also check to see if the bearing is installed squarely by checking around the bearing with the feeler gauges in three or four locations. If it's more than a few thousandths off, square it up.

Install All the Plugs—With the cam bearings in place, you can install the camshaft-bore soft-plug. Also, while you're in the plug-installing business, now's a good time to install the balance of the plugs that go in the block: 6 freeze plugs which fit in the sides of the block—2 or 3 per side—and 6 oil gallery plugs, 3 in front and 3 in back.

Water-Jacket Plugs—The water-jacket plugs are the same as those in the cylinder heads, 1-1/2-in. diameter cup type. It's not necessary to use sealer when installing these, but I recommend it. Install them using the same method used to install the other soft-plugs.

Cam Plug—The rear cam plug is a 2-5/32 in. diameter cup-type plug. Plugs such as this one which must block off oil at the back of the cylinder block must be sealed. Silicone sealer is best. It will ensure that you won't have an oil leak that eventually appears as ugly spots on your driveway, or worse yet, as a well-oiled, slipping clutch, if your car has a manual transmission.

To install the cam plug, use a large-diameter punch—2-in. OD is fine—but don't use a punch smaller than 1/2-in. diameter. Apply a small bead of sealer around the edge of the hole and drive the plug squarely into the block so its plug is even with the inside edge of the hole's chamfered edge. After the plug is in, wipe away the excess sealer to make the job neat.

Front Oil-Gallery Plugs—The front oil-gallery plugs are small 1/2-in. diameter cup-type plugs. You'll need three plugs and a 3/8-in. punch to install them. It will fit inside the plugs. If you don't have a punch this size, use an old 3/8-in. bolt with the end ground off square so it won't distort the plug during the installation. Unlike the rear oil plugs, it's not necessary to use sealer on these plugs. Any leak which occurs will be very small and the oil will end up in the oil pan.

Rear Oil-Gallery Plugs—Due to the need to prevent oil leakage at the back of an engine block, the rear oil-gallery plugs are pipe-thread plugs. If you look through your collection of bits and pieces you should find three 1/4-in. pipe plugs with a 3/8-in. square. If any of your plugs were damaged to the point they can't be re-used, replace all of them with *socket-head* plugs which require an Allen-type hex wrench for tightening. Rather than being made from cast steel like the original plugs, the steel plugs with rolled threads are less likely to gall and seize in the block threads. The socket type also eliminates the problem of rounding the square off. When installing them, regardless of the type you use, coat the threads with sealer and run them in firmly.

Oil-Filter Adapter—If you removed the oil-filter-to-block adapter, reinstall it. You'll need a 1-1/4-in. socket to do this. If you remember from when removing it, the hex-nut that is an integral part of the adapter nut is short, so be careful

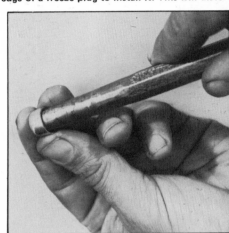

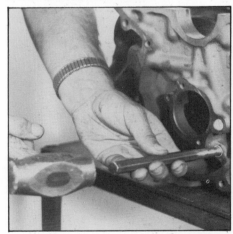

You'll need a large punch to install the front oil-gallery plugs, but not one that fits tight. Otherwise it'll end up trapped after a plug is installed.

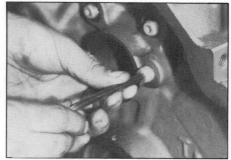

If you're replacing your rear oil-gallery plugs, get a socket-head type rather than the OEM square-head type. It's a 1/4" pipe thread. Install the plugs after coating the threads with the silicone sealer.

Like all the other cup-type plugs, you'll need something large to install the rear camshaft plug. This one is critical because if it leaks you'll have a very difficult time repairing it. Therefore, make sure what you use as a punch comes close to fitting the ID of the plug. Use silicone sealer!

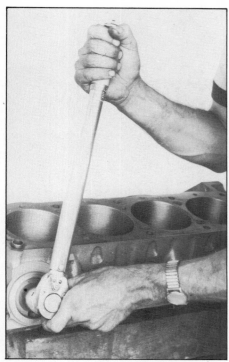

When reinstalling your oil-filter adapter, be sure to push on the socket to keep it from jumping off. Torque it to 80 ft. lbs.

Components which make up a complete camshaft and drive assembly. Center photo shows earlier style counter-sunk thrust-plate bolts and the spacer used with OEM-type cast-iron cam sprockets.

then torquing it in place so the socket doesn't slip off and round off the nut's corners. Torque the nut to 80 ft. lbs.

CAMSHAFT INSTALLATION

How you install your camshaft and prepare it for its first few minutes of initial engine run-in, regardless of whether the cam is new or not, will determine how long it's going to live—30 minutes, 30,000 miles or 100,000 miles? A not-too-uncommon result of an improperly installed camshaft is a lobe/s gets wiped, or rounded off. The damage is not confined to the camshaft because metal particles worn from the lobe and lifter end up well distributed through an engine's oiling system—oil pump, main bearings and everywhere oil is circulated. The oil filter filters out most of the debris, but not all. Not only does this mean the cam and lifters will have to be replaced, it also means replacing all the bearings because metallic particles are imbedded in their soft aluminum-tin or copper-lead overlays. Having to do this immediately after a complete rebuild can make a grown man cry. So, be particularly careful during this part of the engine assembly.

Grease the Camshaft Lobes—Due to the high contact pressures between the cam lobes and their lifters, and the possibility the lifters and lobes will not be receiving much lubrication initially because your engine may have to be cranked a few minutes before it first fires, the cam lobes must be lubricated with something to protect them during these first critical

Applying molydisulphide (MoS$_2$) to the lobes prior to installing the camshaft. Don't skimp here.

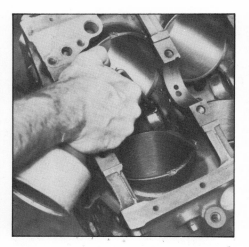

Oil the cam bearings and spread the oil evenly over the bearing surfaces with your finger tips. Carefully slide the cam into place, feeding it in from one bearing to the next. It should fit snugly when the bearing journals engage their bearings.

minutes. This is where molybdenum-disulfide, commonly known as molydisulfide, comes in. Dow Corning's Molykote G-n® paste is readily available in 2.8-oz. tubes, or if you're going to be installing cams for a living you can get it in pint cans and bigger. Valvoline also markets their own molydisulfide—Special Moly EP Grease®.

Before installing the cam in the block wipe the cam bearings clean with lacquer thinner and a paper towel, then oil them well and use your finger tip to wipe the oil around the bearings. Do the same thing with the bearing journals. Apply the moly grease evenly to the cam lobes with a short, stiff-bristled brush or your finger tip and you're ready to install the cam.

Install the Cam—Temporarily install the timing gear onto your cam to provide a "handle." Preferably with your block on its rear face, feed the cam into place being careful not to bang the lobes into the bearings. The lobes and bearings are pretty tough, but there is no point in needlessly damaging something when it is easily avoided. You can install the cam about half the way in the block by supporting the cam by the geared end with one hand and the center bearing with the other. To install it the rest of the way, reach down inside the block with one hand and carefully feed the cam through each bearing web until the journals line up with their bearings. Line the cam up and slide it into place. This is where chamfering the bearings pays off. The cam should fit nicely into place and rotate by hand without much effort. Use the timing gear to turn it with. Take the timing gear off before proceeding with the next step.

Beware of the Thrust Plate and Bolts—There are two things to watch when installing the camshaft thrust plate and its attaching bolts. First, the plate must be positioned right. Your thrust plate will be the idiot-proof type or the correct installed position will be indicated on the plate itself. The plate is grooved to route oil from the front-bearing-journal-to-distributor-shaft oil passage to the camshaft thrust surface and timing chain. This groove must be at *bottom rear* when the thrust plate is installed.

The next items to watch are the thrust plate mounting bolts. You can't just use any old 1/4—20 bolt for this application because of the tight fit between the front of the thrust plate and the back of the cam timing gear. The right bolts have special heads. All 221s used a slotted head, countersunk bolt and all HP289, 302 and 351W engines use a special half-high hex-head, grade-8 bolt. Both are 5/8-in. long. In the intermediate years between 1964 and 1968, the 260 and 289 engines used either the slotted-head or hex-head bolts. To determine which bolts your engine uses look at the thrust plate. If it has countersunk holes, it obviously uses the countersunk bolts, otherwise it uses the hex-head bolts. So make sure you have the right ones or you might end up with the bolt heads jammed against the back of the timing gear.

When you've located the right attaching bolts, install the thrust plate. Remember, it can be installed in four different positions and only one way is correct, providing you don't have the idiot-proof one. It'll work in any position. Snug the bolts up, then torque the slotted head to 8 ft. lbs. or the hex head to 10 ft. lbs.

CRANKSHAFT INSTALLATION

With your camshaft in place you can turn your attention to the crankshaft and its bearings and rear-main oil seal. You'll have to choose the bearings based on the size of your main-bearing journals. The

Early camshaft thrust plates were "idiot-proofed." They didn't have a front, back, up or down and there are grooves in both sides. The later style has to be located a certain way. Its position is indicated so anyone who can read shouldn't have any problem.

Thrust-plate bolts don't have to be torqued much—8 ft. lbs. for the countersunk type and 10 ft. lbs. for the hex head.

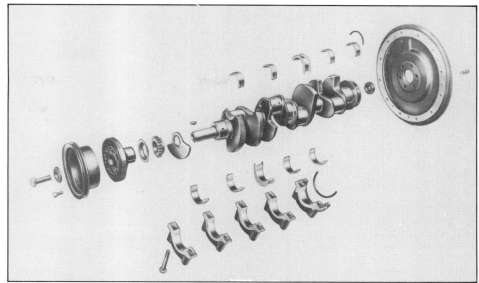
Crankshaft and related components. HP289 uses an additional counterweight which locates behind the timing-chain sprocket with a roll pin. Pilot bearing and flywheel are for standard transmissions only. A flexplate is used with automatic transmissions. Photo courtesy Ford.

Wide flanges of a crankshaft thrust bearing distinguishe it from the single-purpose radial bearing. Thrust bearing locates in the number-3 position in the center.

Sometimes the label on the box bearings come in doesn't tell the truth. Check the bearings to confirm their size. They will be standard (STD) like the one shown or so many thousandths undersized: -010, -020, etc.

rear-main seal comes with the engine's lower-end or over-haul gasket set.

Size the Bearings—Refer to your crankshaft inspection records to determine which size bearings to use—standard or 0.010, 0.020, 0.030, etc. undersize. For example, the standard main-bearing-journal diameter for all small blocks, except for the 351W, is 2.2486 in. with a tolerance range of 2.2482—2.2490 in.

If your main-bearing journals are in this range, you'll obviously use standard bearings. However, if they mike some amount less you'll need the same amount of undersize bearing as the difference from the standard nominal bearing size. Simply put, if a bearing journal mikes 2.2386 in. in diameter, the proper undersize bearing to use is 2.2486 − 2.2386 = 0.0100-in. undersize. So, to determine if and how much of an undersize bearing you need, subtract the diameter of your bearing journals from the standard nominal bearing-journal diameter.

There are two types of rear-main bearing seals you can install in your engine. They are the *rope* and *split-lip* types. The rope seal will probably be included in your gasket set. It consists of two pieces of graphite-impregnated rope. The split-lip seal is made from neoprene, also in two pieces. If you want this seal instead of the rope type, you'll have to lay out another three bucks or so. My reasons for preferring the split seal are the rope seal is more difficult to install. And, the rope seal, as initially installed, creates more friction, meaning the torque required to start the crankshaft turning will be approximately 10 ft. lbs. more. However, the relative effectiveness of the two seals is the same for street operation and the initial high friction of the rope seal drops off considerably during the first few hours of engine operation to the point where there shouldn't be any distinguishable difference between the two types of seals.

Check Bearing-to-Journal Clearance—You have already determined the size bearings to use by measuring the bearing journals with a micrometer. Due to the accuracy of the mikes and the close tolerances to which bearings are manufactured, direct clearance checks are unnecessary, assuming the right bearings got in the right boxes. However, we all know that assuming can cause considerable trouble and checking just takes time.

The most simplified clearance check is to install the crankshaft in its bearings in the block and turn it by hand to see if it rotates freely. The other, more involved method of checking clearances is with Plastigage. Because both methods are done in the normal crankshaft installation sequence let's get on with the installation and save some words in the process.

Install the Main-Bearing Inserts—With the engine block on its back, wipe the bearing bores clean so there won't be any dirt trapped between the bearing inserts and their bores after they are installed. The same thing goes for the bearings. Wipe them clean using *paper towels* and a solvent such as lacquer thinner. Clean both their front and back sides. The actual bearing surface will be coated with a white residue as it comes out of the box. It's OK, just wipe the bearing so you're certain the working surface is clean.

Each bearing half has a bent down tab at one end. The top bearings, the ones

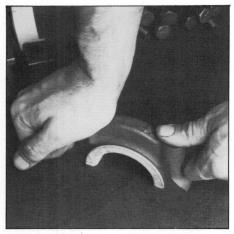

Before installing the bearing inserts, wipe the bearing bores clean as well as the back of the inserts. Bearings with grooves and oil holes go in the block and the plain ones go in the caps. Thrust bearings requires a little extra push to get them seated.

that go in the block, are grooved. Their tabs are located off-center. The bottom halves aren't grooved and their tabs are centered. Locating the tabs in this manner helps prevent the bearings from being installed wrong. Install the bearings in the block by first fitting the tab of each bearing in the notch at the edge of the bearing bore. Fit this end of the bearing flush with the edge of the bearing bore. Hold it in this position with a finger or thumb while you force the bearing into place by pushing down on the opposite end with your other thumb. All the bearing halves will go in with little effort except for the center one. It also serves as a thrust bearing, and will require more force because the thrust flanges fit tightly around the bearing web.

Before installing the other bearing halves in the caps, lightly file the cap-to-block mating surfaces to ensure the caps will fit properly in their registers. To do this you'll need a fine-tooth, large flat file. Lay the file on your bench and stand a cap up on the file. Lightly run the cap over the file a few times when holding the cap square against the file. Be careful not to remove any material from the cap except for nicks or burrs—small projections which are raised above the normal surface of the cap. After doing this install the non-grooved bearing halves into the caps.

Rear-Main-Bearing Seals—If you have the split-lip type rear-main seal, you can install it now providing you aren't clearance checking with Plastigage. However, if you have the rope type, you'll have to wait till later if you want to check main-bearing clearances. With this in mind let's get on with installing both types of seals.

Split-Lip-Seal Installation—If your engine was originally equipped with a rope seal—

it was as it came from the factory—there will be a sharp pin centered in the seal groove in the rear-main cap—it'll have to come out. To remove the pin use a small punch to drive it out. Presumably you've already installed a bearing in the cap, so remove it now so it won't get damaged and reinstall it after you have the seal in place. To remove the pin, place the cap down over the end of a 1-in.-wide piece of wood—so when you are driving the pin down out of the seal groove, the load taken by the cap is directly below the pin. Otherwise, the cap can be sprung, making it unsuitable for use.

With the rope-seal locating pin out of the main cap fill the pin hole with silicone sealer, install the seal and reinstall the bearing insert. Lightly coat the seal halves with oil, and while installing them don't let oil get on the block and cap mating surfaces. If it does, wipe it off with a paper towel and solvent. *The lip of the seal must point toward the front of the engine when installed.* Also, don't line the ends of the seal halves up with the cap and block mating surfaces. Rotate the seal halves in the block and cap so one end projects approximately 3/8 in. up from the block and cap mating surfaces, and so the seal will fit together within the block and cap when the cap is fitted to the block.

Rope-Seal Installation—Put the seal in the block first. Lay the rope in the groove edgeways. Push it in with your thumb, leaving both ends of the seal extending above the block and cap mating surfaces. With the seal in place, work it into the groove with a cylinder—I use a 1-1/2-in. socket—and a hammer to tap on it as you work the cylinder back and forth in a rolling motion. You don't have to pound it in, just use a light tapping motion. When it appears the seal has filled the groove,

A large socket and a brass hammer being used to install a rope seal in the block. Work around the seal until it is all the way down in its groove, then trim the ends flush with the block.

trim the excess ends off flush with the block using a sharp knife or a single-edge razor blade.

Now you can install the other half in the bearing cap. It'll be more difficult to handle just because the cap will be clumsy to handle while you're installing the seal. It's a whole lot easier if you cradle the cap in something like a vise. *Don't clamp the cap in a vise.* Just open the jaws far enough to permit the cap to sit square. You can then install the seal as you did in the block. When the seal is bottomed in its groove, trim the ends off flush.

Get Your Crankshaft Out of Storage—With the main bearings in place you are now ready to drop your crankshaft in the block. If you are using the rope-type rear-main-bearing seal and are going to check bearing clearances, or have the split-lip-type seal and are going to check bearing clearances using Plastigage, don't install the crankshaft with the seals. Anyway, you can now reintroduce your crankshaft to daylight. Use solvent and paper towels to clean all of the bearing journals of their

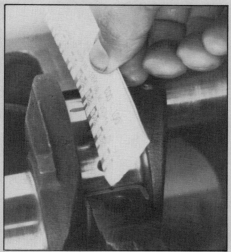

Using Plastigage to check bearing clearance. Lay a strip of Plastigage the full length of the bearing journal, then install the bearing and cap and torque bolts to specification. Make sure *all* bearing journals are free from oil and *all* the other bearing caps are torqued in place first.

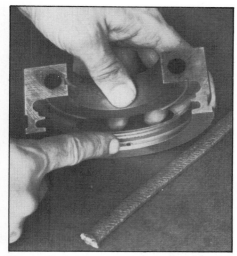

Sharp pin in the rear main-bearing-cap seal groove prevents the rope seal from rotating—don't remove it. If you're installing a split-lip seal, remove the pin and fill the hole with silicone sealer to prevent a leak.

CHECKING JOURNAL-TO-BEARING CLEARANCES
ROTATING METHOD

Generally the simplest method of confirming you've installed the correct size bearings for a given size bearing journal is, first rotate the journal in the bearing or vice versa to check for insufficient clearance. If the rotation is free it's OK. To check for excessive clearance try to rotate it 90° to the normal direction of rotation. If some play, or movement is felt, the clearance is too much. Let's look at how this is applied to crank and rod bearings.

Crankshaft Bearings—Install the crankshaft using the methods outlined in this chapter. If you are using a rope-type oil seal, don't install it until after you've checked the bearings. With the crankshaft in place, the bearing caps installed and their bolts torqued to specification, rotate the crankshaft by hand. If it rotates freely the bearing clearances are enough. To check for excessive bearing clearance, lift the crankshaft by its nose. Try to wiggle it up and down. If you don't feel any movement, or play, clearance is not excessive and you can proceed with your engine buildup.

Connecting-Rod Bearings—Connecting-rod bearings can be checked with the crankshaft out of the block. This is because the pistons are not installed in their bores for the check. To check the rod bearings you'll first have to install the rods on their journals in their normal positions, one at a time. Oil the bearings well with motor oil. After torquing the nuts to specification rotate the rod back and forth on the journal. If it's free there is enough clearance. Now wiggle it sideways, or at least try to. If there's no play then the clearance is not excessive and the bearings are OK.

PLASTIGAGE METHOD

Plastigage is a strip of wax which, when installed between a bearing and its journal, flattens or squeezes out to a width inversely proportional to the clearance between the journal and the bearing. Plastigage comes in a paper sleeve with a printed scale to measure the flattened Plastigage. It reads out directly in thousandths of an inch bearing-to-journal clearance. You bought green Plastigage to measure 0.001—0.003-in. clearance range.

To use Plastigage, cut a length of it which coincides with the bearing width. The bearing and journal must be free from any oil because Plastigage is oil-soluble, and any oil will cause a false reading. Lay the Plastigage on the top of the bearing journal or centered in the bearing, and carefully install the cap with the crankshaft laying in the block on its new bearings, but without its rear-main oil seal, or the piston and rod assembly positioned against its bearing journal. Torque the bolts or nuts to specification. Be careful not to rotate the bearing and journal relative to each other, otherwise the Plastigage will smear and you'll have to remove the cap and start over again. After you've finished torquing the cap, remove it and measure the bearing clearance by comparing the squeezed wax width with the printed scale on the Plastigage sleeve.

It's not necessary to check all the bearings unless you want to be particularly careful. But one thing for sure, you can't be faulted for checking to make certain.

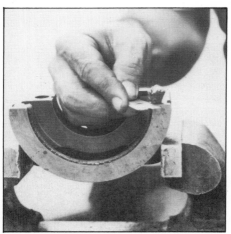

Other half of rope seal is installed in cap as it was in the block, using a large-diameter socket and mallet. Cap is not clamped in the vise, it's just cradled. Trim the seal ends even with the cap.

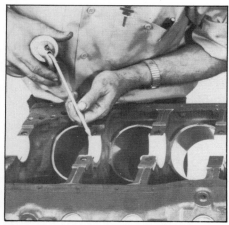
Prior to installing the crankshaft, wipe main bearings with a paper towel soaked in lacquer thinner to remove the white residue and any dirt which may be deposited on them. Spread oil around the bearings to coat them—oil the seal too.

Carefully lower the crankshaft into place. Watch your fingers at the rear seal. You'll need a finger in the end of the crankshaft to support it.

protective oil coating. They will have collected some dust and dirt particles by now and spray-type oil is not suitable for engine assembly anyway.

If you are installing your crankshaft to stay, check its oil holes again for dirt. This is your last chance to catch any contaminants that will otherwise be pumped in oil directly to the bearings and journals the instant your engine is started. Oil the bearings in the block as well as those in the caps. Also spread some oil on the seal. Lower the crank carefully into the block and give it a spin once it's seated on the bearings. Now you're ready for the caps. With their bolts lightly oiled and loosely installed in the caps, position them on their bearing journals using the cast-in numbers and arrows to determine their position and direction. The arrow should point toward the front of the engine and the number will be its position. The rear-main cap won't have an arrow or number. Its position and direction are obvious, but the others are not.

Seal the Rear-Main Cap—For final crankshaft installation, the rear-main-bearing cap and cylinder block parting line must be sealed, otherwise it will leak. Run a small bead of silicone sealer—about 1/16-inch wide—in the corners of the cylinder-block's rear-main-bearing register from the back edge of the block to even with the front edge of the crankshaft oil-slinger groove. Also, apply another 1/16-inch wide strip of sealer to the cap in from both edges of the cap and in line with the seal. Stop about 1/8 inch short of the seal. Refer to the sketch for a pictorial explanation of applying the sealer. Oil the seal before installing the cap.

Don't Tighten the Main-Bearing Caps Down Yet—Before threading the bearing-cap bolts in the block, the caps must be located in their registers. If the bolts are

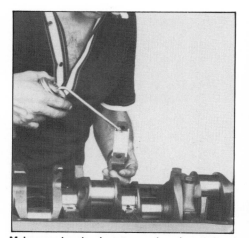 
Make sure bearing inserts are clean by wiping them off, then coat them with oil. Do the same to the crankshaft main-bearing journals. Oil the bolt threads of the cap bolts. Seat the bearing caps in their registers. Locate one side of the cap in its register, then tap the opposite side into place by hitting the side of the cap at about a 45° angle.

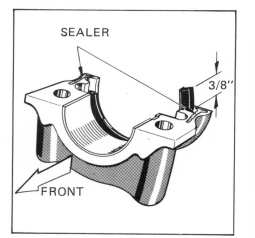 
Drawing shows how to position a split-lip-type seal. It should project approximately 3/8 in. out of the block and the same amount from the opposite side of the bearing cap so the two will engage when the cap is installed. Apply a small bead of silicone as shown just prior to installing the cap. In the other photo I'm applying some sealer to the ends of a rope seal prior to installing the cap.

tightened in an attempt to draw the caps into place, you may end up ruining your block and some of the bearing caps. The edges of the bearing register or caps may end up being broken off. Locate one edge of each cap against its register, then tap the cap with a hammer on its opposite side at about a 45° angle while pushing down on the cap. It should snap into place. After the caps are in place you can run the bolts down, but don't tighten them yet.

If you have the rope-type seal, now is the time to see how much torque is required to turn your crankshaft against its friction. Tighten down the rear-main cap only. Do each bolt progressively. Don't run one bolt down, snug it up and torque it, then go to the other one. The right way to do this is to run them both down, snug one, then the other and torque them both a little at a time until both bolts are torqued 60–70 ft. lbs. for all engines but the 351W. If you have a Boss 302 don't install the outer main-cap bolts yet. Torque 351W main-bearing bolts 95–105 ft. lbs. Now, install the damper attaching bolt and washer so you can turn the crankshaft with your torque wrench. Expect 10–15 ft. lbs. to turn it. Now, you'll know one of the major causes of the high effort required to turn your crankshaft during the rest of the engine assembly.

Seat the Thrust Bearing—Using the same technique you used for torquing the rear-main-cap bolts, secure the others except for the center cap—the one with the thrust bearing. Tighten its bolts finger-tight, then line up the bearing thrust faces by forcing the crankshaft back and forth in its bearings. Use a couple of big screwdrivers to pry between two other bearing caps and crankshaft throws and counterweights. After doing this a few times, pry the crankshaft forward. Hold it in this position while you torque the two bolts which secure the thrust-bearing cap. Check crankshaft turning torque now and you should find a small increase.

Crankshaft End-Play—Your crankshaft end-play should be 0.004–0.008 in. with a 0.012-in. maximum. To check end-play accurately, a dial indicator is the best. However, if you don't have one, feeler gauges will work. To use the feeler gauges, pry the crankshaft in one direction, then measure between the crankshaft's thrust face and bearing. However, if you are among the affluent and possess a dial indicator, mount it with its plunger parallel to the crankshaft center line and with the plunger tip firmly against a flat surface on the crankshaft's nose or flywheel mounting flange. Pry the crank away from the indicator, then zero the dial gauge. Now, pry the crank in the other direction to take up its end-play. Read total end-play directly. Repeat this a few times for verification.

I'm torquing the rear-main cap down first to check the friction caused by the rope seal. Using the damper bolt and washer installed in the nose of the crank, torque due to friction measured 13 ft. lbs.

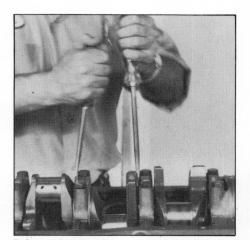 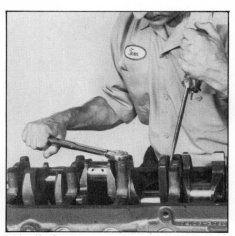

Before tightening the center main-bearing-cap bolts, seat the bearing inserts first. Wiggle the crankshaft back-and-forth, then hold it forward while you torque the bolts: 95–105 ft. lbs. for the 351W and 60–70 ft. lbs. for the others.

If, by rare chance, end-play is under the 0.004-in. minimum, the thrust-bearing flanges will have to be thinned, meaning the crankshaft must be removed! Thin the thrust bearing's front flange by laying some 320 grit sandpaper on a *flat* surface and *lapping* away as much bearing material as required to bring end-play within tolerance. Before doing any sanding, use a micrometer to measure the width across the bearing flanges. Check it at both ends and in the middle. Do both bearing halves and record the initial results. While sanding the bearings, hold the bearing straight up and avoid any rocking motion. Periodically measure the bearing to make sure you are sanding the material evenly and not excessively. When you've matched both bearing halves, clean and reinstall them and the crankshaft following the correct procedure. Check end-play again to confirm what you did had the right effect.

If the crank's end-play exceeds 0.012 in., trade your crankshaft in on a crankshaft kit and let the reconditioner worry about the worn one. Having too much end clearance is highly unlikely considering you've inspected your crankshaft and declared it sound or have had it reconditioned or replaced. However, unlikely or not, end-play is one of those things which must be checked regardless of the consequences. Like the man says, "pay now or pay later." What you'll have to pay now will be a fraction of the cost later.

PISTON AND CONNECTING-ROD INSTALLATION

If you've yet to assemble your pistons and connecting rods, refer back to page 70 for how it's done. However, if they are assembled you can prepare them for installation into your engine. The equipment you'll need includes a small flat file, a ring compressor, hammer, a large tomato can with about a half quart of clean motor

Checking crankshaft end-play on this Boss 302. Although less accurate, the same thing can be done with feeler gauges by taking up the slack in end-play in one direction and measuring between the bearing insert and the crankshaft thrust faces. Photo courtesy Ford.

Check your ring end gaps before installing the rings on their pistons. Square each ring in its bore like I'm doing with this old piston. Use your feeler gauges to check end gap.

oil in it, an oil can and some bearing-journal protectors to slip over the rod bolts. You should consider a ring expander too, even though it's possible to install the rings without one. The expander ensures the pistons won't get scratched or the rings bent during the installation. If you can't readily pick up a couple of bearing-journal protectors at your automotive-parts store, two 2-inch lengths of 3/8-in. rubber fuel hose will suffice.

Gap Your Rings—Piston rings are manufactured to close tolerances to fit specific bore sizes, however, you should check their end-gaps just the same. Both compression rings should have a 0.010—0.020-inch gap, or more precisely, 0.004-inch gap for each inch of bore diamter. For example, the compression rings for a 4-inch bore should be gapped at 0.004 x 4 = 0.016 in. However, if your ring gaps are in the 0.010—0.020-in. range, they'll be all right. Gap range for the two oil-ring side rails is much wider at 0.015—0.055 in.

To check a piston ring's end gap, install it in a cylinder bore. Carefully fit the ring in the bore and square it up using an old piston or a tomato can. There's no need to push the ring down the bore more than a half-inch unless your block hasn't been rebored. In this case you should push the ring down the bore to the bottom of its travel. The reason is taper in a cylinder causes a ring to close near the bottom of the stroke and minimum end-gap clearance is what should be checked to prevent the ends of a ring from butting when the engine is at maximum operating temperature. If any of the gaps are too large, return the complete set of rings for replacement. If a gap is too small, it can be corrected by filing or grinding the end of the ring.

When filing or grinding a piston ring the first rule is to file or grind the end of the ring with the motion *from its outside edge toward its inside edge*. This is particularly important with moly rings. The moly filling on the outside periphery will chip off from the outside edge. When filing, don't hold the ring and file in your hands, clamp the file in a vise or to a bench with a C-clamp or Vise-Grip pliers and move the ring against the file. You'll have better control using this method. For grinding a ring, use a thin grinding disc, one that will fit between the two ends. Just make certain the rotation of the disc is in the right direction. Also, be sure the ring is securely held so you don't break the ring as you grind it.

Break any sharp edges from each piston-ring end that you filed or ground using a *very fine* toothed file or some 400-grit sandpaper. The inside edge will have the biggest burr, the outside edge the least—if it has any at all—and the sides will be somewhere in between. Just touch the edges with your file or sandpaper. Remove only enough material so you can't feel any sharp edges with your finger tip.

When you've finished one complete ring set, keep it with the piston and rod assembly that's to go in the bore you used for checking their end-gaps. Repeat this procedure until you've checked and fitted all the rings. Also, it's best to keep

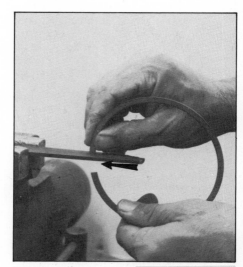
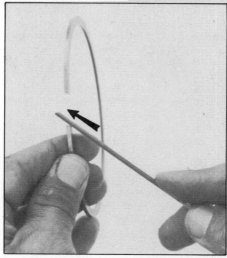

To increase ring end gap, file one end of the ring with a fine-toothed file. With the file firmly clamped, move the ring against the file in the direction of the arrow. Lightly touch the corners of the end you filed to remove any burrs which may have developed.

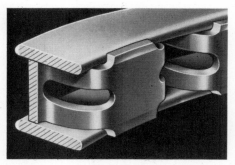
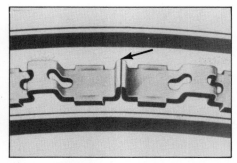

Section of an oil-ring assembly. Notice how the expander fits behind the two rails to spring-load them against the cylinder bore. The ends of a correctly installed expander must butt (arrow). Drawings courtesy Sealed Power Corporation.

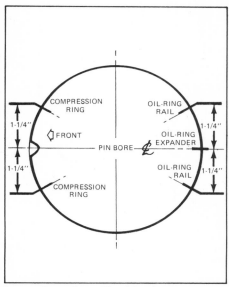

Locate ring end gaps like this prior to compressing them for installing the piston-and-rod assemblies into your engine.

the rings in the envelopes they came in. The ring manufacturer usually organizes the rings in sets with its rings in order—top compression ring, second compression ring and oil-ring assembly. Be careful not to get the rings mixed up, particularly the compression rings. The differences between these two rings are not readily apparent, so it's best to keep them in order to avoid installing them on the piston in the wrong order.

Install the Rings on the Pistons—With the rings gapped, you're ready to install the rings on their pistons. This can be done without any special tools, but I suggest using a *ring expander*. It will greatly reduce the possibility of ring or piston damage during the installation. An expander costs much less than another set of rings.

So you won't have to horse around trying to hold a piston still while trying to install its rings, lightly clamp each rod and piston assembly so your hands will be free to concentrate on holding the ring you are installing. Remember, don't clamp the rods very tight, particularly in a vise. They are easily bent. Clamp the rods between two small blocks of wood with the rod far enough down in the vise so the piston can't rock back-and-forth. Be careful not to clamp on the piston.

Oil Rings—When installing rings on a piston, start with the oil ring and work up. You won't need the ring-expanding tool for the oil-ring components—the expander ring and the two side rails. However, you can save yourself some time later on by installing the three rings which make up the oil ring so their end-gaps are in the right relationship to one another rather than waiting until time to install the piston and rod assembly in your engine. The idea here is to *stagger* the gaps so they don't coincide with each other. If this were to happen the oil ring would pass excessive amounts of oil.

The first thing to do is locate the front of the piston as referenced by its notch or the arrow which is stamped in its dome. Generally, if it's a cast piston it'll have the notch, and an arrow with a forged one. Use the nearby sketch for positioning the end-gaps for all the rings. With the front of the piston located, install the ring expander with its gap directly at the front of the piston. Its job is to *expand*, or spring load the two side-rails against the cylinder bore for sealing. Be careful when installing this ring to make sure its ends butt and don't overlap. Now you're ready for the side-rails. Insert one end in the oil-ring groove on top of the expander. Hold the free end of the ring with one hand while running the thumb of the other hand on top of the ring and around the piston while pushing it into place. Don't let the free end of the ring dig into the piston as you bring it down over the piston and into its groove. Install the other side-rail below the expander ring using the same method. Position the rail gaps as shown in the drawing.

Compression Rings—Compression rings are much simpler in design, but more difficult to install than oil rings because of their rigidity. They are easily damaged, and can damage the piston during installation if done incorrectly. For instance, as opposed to an oil ring, which has no up or down, a compression ring usually does. So, keep your eyes open when installing these guys. Look for the "pip" mark—a little dot or indentation—which indicates the top side of a ring. The pip mark must be up when the ring is in its installed position, otherwise the ring's twist will be in the wrong direction resulting in bad things happening, things like excessive blowby and ring and piston-groove side-wear. *Twist* is a term which describes how a ring bends, or the shape it takes in its groove. Twist is used for both oil control and sealing combustion pressures. Compression rings with no twist also have no pip marks. If you happen to get a set like this, you can install them either way, but inspect the rings to make absolutely sure. Also, read the instructions which accom-

After the expander, the top oil-ring rail goes on next. Hold the free end of the rail so it doesn't scratch the piston and position its end gap now.

panied the rings and follow them *exactly*.

The difficult thing about installing a compression ring is spreading it so it'll fit over the piston while not twisting or breaking the ring or gouging the piston with the ends of the ring. This is why I highly recommend using a ring expander.

There are different styles of ring expanders. The top-of-the-line type has a circular channel for the ring to lay in while it is expanded for installation. The less-exotic expander doesn't have this channel. With this type, one hand is used for expanding the ring and the other for controlling the ring. Regardless of the type you have, don't spread the ring any more than necessary to install it. Install the second (lower) compression ring

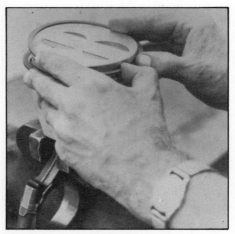

Two ways of installing a compression ring—with or without a ring expander. If you elect not to use a ring expander, be careful. It's easy to bend or break a ring. If this happens you could've saved money by purchasing and using an expander. Regardless of how you do it, make sure the pip marks are up.

Tools for installing piston-and-connecting-rod assemblies. Bearing-journal protectors for the rod bolts, a ring compressor, a wooden-handle hammer for pushing the piston down into its bore and oil in a squirt can and tomato can for oiling the piston and rod bearings.

first. Make sure the pip mark is up, expand the ring while making sure it doesn't twist and install it over its piston into the second groove. Do exactly the same with the top ring, installing it in the top groove. The end-gaps on these two rings should be placed as shown in the drawing. They should be checked for position *immediately before* being installed in the engine.

You may decide to install your compression rings without the aid of a ring expander, so I'll tell you how to do it right. Wrap the ends of your thumbs with tape. The sharp ends of a compression ring really dig in when you're spreading it apart. Position the second compression ring over the piston with the pip up. Rotate the ring over the edge of the piston so its ends are lower on the piston than the rest of the ring. With your thumbs on the ends of the ring so you can spread it apart and your fingers to the side for controlling it, spread the ring apart and rotate the ring over the piston with the end gap over the groove. When the ring lines up with the groove all the way around, release it into its groove. When your thumbs revive, install the top ring using the same method. It'll be easier because it won't have to go down over the piston as far.

PISTON AND CONNECTING-ROD INSTALLATION

For installing piston-and-connecting rod assemblies into your engine you must have a ring compressor, two bearing-journal protectors for the rod bolts, an oil can, something to push the pistons into their bores—a hammer will do a good job—and a large tomato can containing a couple of inches of motor oil. You don't have to have the last item, but it is the most convenient way I know to pre-lube your rings and pistons prior to installing them.

Get Everything Ready—Just like every other facet of engine building, you have to be organized when installing connecting rods and pistons. All the tools must be within reach, everything must be clean and the engine positioned so you'll be able to insert each piston-and-rod assembly into its bore while guiding the rod into engagement with its crankshaft bearing journal.

Before starting the actual piston and rod installation, clean your engine's cylinder bores of their protective oil coating. Use paper towels for this, not rags. Look your pistons and rods over too. Regardless of how clean the rest of the assembly is, wipe the bearing inserts to be sure they are clean. If you're doing the engine assembly on a bench, roll the block upside-down so one of the deck surfaces hangs over the edge of the bench. This will give you clear access to the top and bottom of the bores and to the crankshaft bearing journals. Have the pistons close by and organized according to their position in the engine. This way you won't have to hunt for each piston even though it gets easier as you go along. With all the tools ready you should be ready to slide the pistons in their holes.

Crankshaft Throws At BDC—If you've removed the crankshaft damper bolt and washer, replace them. You'll need them so you can turn the crank periodically during the rod-and-piston installation process. The throw, or connecting-rod bearing journal, must be lined up BDC for the cylinder that's to receive a piston and rod. This is so you'll have room between the bottom of the cylinder and the crank journal to guide the big end of the connecting rod into engagement with its journal as you slide the piston down the bore.

Once you have the crankshaft in position, remove the bearing cap from the rod you're going to install. Be careful not to knock the bearing inserts loose from the rod or cap if you installed them earlier for checking journal-to-bearing clearance with Plasti-Gage®. Above all, don't mix them with other inserts. The best way of avoiding this is only remove one rod cap at a time. Slip the journal protectors over the rod bolts. Liberally oil the piston rings and ring grooves and the skirt. Here's where the big tomato can with oil in it comes in handy. Just immerse the top of the piston in the oil, then spread the oil all over its skirt. While you're at it, oil the wrist pin and the bearing inserts—both in the cap and the rod. Spread the oil evenly on the bearings with your finger tip.

Ring compressors vary in design from one to another, so follow the directions that accompanied yours. However, if you don't have the directions for one reason or another, here's how you use one. First, there are two basic styles. Most common and least expensive operates with an Allen wrench. The compressed position is held by a ratchet which is released by a lever on the side of the compressor sleeve. The second type uses a plier arrangement to operate the compressor. A ratchet on the plier handle holds the rings in the compressed position. This type is available from Sealed Power Corporation with sleeves or bands to cover pistons sizes 2-7/8–4-3/8 in. diameter.

To use a ring compressor, fit it loosely around the piston with the long side of the sleeve pointing up in relation to the piston. With the clamping portion of the compressor centered on the three rings compress the rings while wiggling the compressor slightly. This helps the compressor to compress the rings. When the compressor is snugged tightly around the piston, put the protective sleeves over the rod bolts.

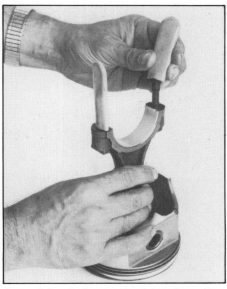  

After wiping the bearings and their bores clean, install them, then slip thread protectors over the rod bolts. Oil the bearings as well as the piston by submerging them in oil.

After compressing rings, guide connecting rod and piston into its bore and in the right direction. Check connecting-rod number and arrow or notch in top of piston. It must point to the front of the engine. Push or lightly tap piston into its bore while guiding connecting rod onto its bearing journal. If you feel any slight hangup, STOP! Pull the piston out and start over again after recompressing the rings.

Using the notch or arrow on the piston dome or the rod number as reference for positioning the assembly, insert the piston in its bore while being careful not to let the connecting rod bang around in the bore. The lower portions of the piston skirt should project out the bottom of the compressor sleeve. Insert the piston in the cylinder bore and push down on the compressor to square it up to the block's deck surface. With your hammer, using its head as a handle, push or tap lightly on the head of the piston to start it down the bore. If the piston hangs up before it's all the way in, **STOP**. Don't try to force it in the rest of the way. What's happened is a ring has popped out of the bottom of the compressor before entering the bore, consequently if you try forcing the piston any farther you'll probably break a ring land, resulting in a junked piston. Release the compressor and start over. It's no big loss, just a couple of minutes rather than a piston and a ring. When the last ring leaves the compressor and enters its bore, the compressor will be free of the piston. Set it aside and finish installing the piston and rod.

Using one hand to push on the piston and the other to guide the rod into engagement with its rod journal by sliding the piston down the bore, guide the rod bolts so they straddle the journal, then tap lightly on the piston dome until the rod bearing is firmly seated against the journal. Remove the sleeves from the rod bolts.

You can now install the bearing cap. Make sure its number coincides with the one on the rod and is on the same side. Install the attaching nuts and torque them to specification, 19–24 ft. lbs. for 221, 260, 289 and 302 engines and 40–45 ft. lbs. for the HP289, Boss 302 and 351W engines.

Now that the first rod and piston are installed, turn the crankshaft so its throw lines up in the BDC position for the next cylinder. Install its rod and piston assembly as well as the remaining two in this bank, remembering to put the protective sleeves over the rod bolts each time. When finished with this side, roll your engine over and install the last four. With that you can take a few steps back and take a good look at your engine—it's starting to look like one now.

Boss 302 Windage Tray—If you have a Boss 302 you can now install the *outer* main-bearing-cap bolts and its *windage tray*, or oil-baffle under these bolts. Torque them 35–40 ft. lbs.

TIMING CHAIN AND SPROCKETS

Before you can install the timing chain and its *sprockets* (often referred to as *gears*), you'll have to rotate the crankshaft and camshaft to their cylinder 1 firing positions. For the crank, just bring piston 1 to the top of its bore. The keyway in the crankshaft extension will be pointing up in relation to the block. Position it a little left of straight up, looking at the keyway from the *front* of the block. You'll see why later. The position for the cam won't be as apparent. Position it so its

A simple check to make sure pistons are in the right holes and in the right direction—arrows or notches pointing forward and numbers pointing toward their cylinder bank.

Torque the cap bolts now; 40—45 ft.lbs. for the HP289, Boss 302 and 351W and 19—24 ft.lbs. for all others.

sprocket drive-pin is directly in line with the crank's centerline, between the cam and the crank.

Don't Forget the Key—If the key is missing from the crankshaft keyway, find it and install it. It must be there to index and drive the cam. It also drives the accessory-drive pulley. Before laying the key in place, feel the edge of the keyway. If it has a little burr or a raised edge, file it flush with the crankshaft's normal surface, then install the key.

Install the Crankshaft Sprocket First—If the damper bolt and washer are still on the end of your crankshaft, remove them so you can install the crankshaft timing-chain sprocket. Don't slide it all the way on, just enough so its keyway slot begins to engage the key. Notice the timing mark on the sprocket pointing straight up relative to the engine.

Before installing the cam sprocket on the cam, find its timing mark. It'll be on the front face of the sprocket between two of the teeth and in line with the drive-pin hole. This mark must point down, lining up with the crank sprocket when cylinder 1 is on TDC. Using the mark as reference, hold the sprocket in the position it will go on the cam and drape the timing chain over it. Bring the sprocket and chain up to the engine and mate the chain to the crank sprocket so the timing marks on the two sprockets line up—are directly opposite one another. While lifting up on the cam sprocket to keep the chain tight, slide the crankshaft sprocket back simultaneously with the cam sprocket and chain until the cam sprocket engages the cam nose and drive-pin. You won't be able to slide it on all the way because the drive-pin hole in the sprocket should be to the right of the drive-pin because of

the position you left the crank in—to the left, or counter-clockwise. To bring the drive-pin and drive-pin hole in line with each other, reinstall the damper bolt and washer in the end of the crank and bump the crank in the clockwise direction while pushing on the cam sprocket. When it's lined up you'll feel it engage the drive-pin. You can now slide both sprockets all the way on. Before you go any farther, make sure both sprocket timing marks point straight at each other. If they aren't you'll have to back both sprockets off until the cam sprocket is loose and jump the chain a tooth in the right direction and reinstall the chain and sprockets and recheck them.

Install the Fuel-Pump Cam—Secure the camshaft timing-chain sprocket, but don't forget the fuel-pump cam. It goes between the attaching-bolt and washer, and the sprocket. There are two holes in the cam, a big one and a small one. Locate the cam with the small hole over the end of the cam drive-pin which protrudes out of the sprocket and the large hole over the center of the sprocket and the cam. Install the attaching bolt—3/8-24 x 1.50-in. long, grade 8—washer and torque the bolt to 40—45 ft. lbs.—a dab of Loctite® on the bolt threads is good insurance.

Slide the oil slinger over the end of the crankshaft extension and you're ready to install the timing chain cover. Don't worry about the slinger being loose. When the crankshaft damper is installed, it will sandwich the slinger between it and the crankshaft timing-chain sprocket. Just make sure the slinger is on in the right direction—so the top of the hat is *against the sprocket*—and engaged with the key.

TIMING-CHAIN COVER AND CRANKSHAFT DAMPER

If you haven't cleaned the front engine

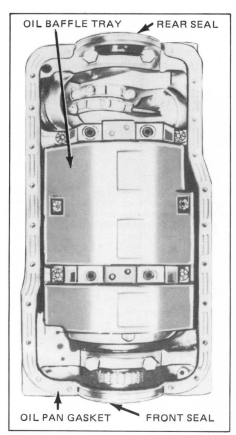

Boss 302 windage tray can be installed after connecting-rod nuts are torqued. Install it under number-2 and -3 outer main-bearing cap bolts. Torque them 35—40 ft.lbs. Photo courtesy Ford.

With camshaft dowel pin positioned at the bottom and crankshaft key slightly left of straight up looking at the front of the engine, loose-assemble timing chain and sprockets timing marks pointing at each other. Start the bottom sprocket on the crank, then slide the complete assembly into engagement with the crankshaft key and fit the cam sprocket over the end of the camshaft.

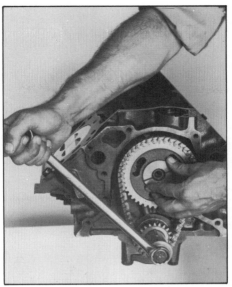
To engage the camshaft sprocket with its dowel pin, turn the crankshaft clockwise using the damper bolt, washer and a wrench while pushing back on the cam sprocket at the same time.

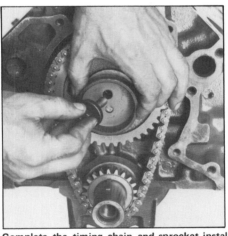
Complete the timing chain and sprocket installation by installing fuel-pump actuating cam and its attaching bolt and washer to camshaft end. Small hole in fuel-pump cam locates over dowel pin with bolt going in the other one. Torque bolt to 40—45 ft.lbs. Slip crankshaft oil slinger against the crank sprocket and you're ready for the front cover.

and installing. With the cover supported directly in front of the seal bore by laying it on a firm flat surface, locate the seal in the cover and over its bore so its lip will point toward the engine when the cover is installed. It's best to use a cylinder that's a little smaller than the outside diameter of the seal, but will seat against its edge for driving it in. This lessens the possibility of distorting the seal during the installation. I used a very large diameter socket—one which you may not have. If need be, you can install the seal by tapping it in around its edge a little at a time while being very careful not to cock it in its bore. Regardless of the method used, the seal is fully installed when it bottoms against the backside of the cover's front lip. When it's bottomed all the way around, the cover is ready to be installed on the engine.

Prepare the Damper—Nothing accurately positions the front cover to the front of the engine block. Consequently the crankshaft seal needs to be accurately located, or centered on the crankshaft while it's being installed, otherwise it'll end up off-center, causing uneven seal loading and a probable oil leak. To center the front cover you'll need a special tool or the obvious thing, your crankshaft damper. If you haven't cleaned it up yet, do so now. The main thing is to inspect the rubber bond between the outer ring and the inner

cover yet, do it now. Remove all the dirt and grease and all the old gasket material from the sealing surfaces. Use a punch and a hammer to remove the crankshaft front oil seal. Carefully knock the seal out the back of the cover by placing the punch behind the cover's inside lip and gradually work around the seal until it falls out. Don't try to knock the seal out with one whack or you may succeed in breaking the cover. With the old seal out of the way, clean its bore in the cover.

Install the New Seal—You'll find a new front crankshaft oil seal in your gasket set. Coat the outside periphery of its shell with some grease. This will assist in sealing

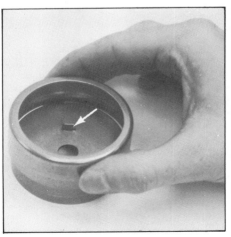

Two-piece fuel-pump cam is retained the same as the one-piece cam, however it locates with a bent-down tab (arrow) in the cam-sprocket hole on top of a shorter dowel pin. Outer ring turns on the inner cup as can be seen by the installation.

Remove front engine-cover seal with a punch and hammer by working through the cover opening. Be careful because the cover is easily broken.

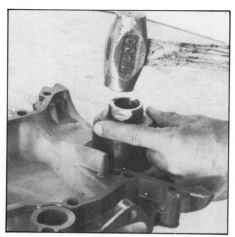
Support cover on a firm flat surface and drive front-cover seal into place using a very large socket which fits seal's outer shell.

Not so much for the installation, anti-seize compound is used in the event the damper has to come off again.

This little sleeve can save you considerable money. It presses onto the damper over a damaged or badly grooved seal surface.

one. Try to force them apart longitudinally, or back and forth. If there is noticeable movement the bond has failed and the damper should be replaced. Also inspect the inner bore and the sealing for any roughness, particularly the seal surface. Smooth out any imperfections with some 400-grit emery cloth. It's also a good idea to give the sealing surface the same treatment you gave the crankshaft bearing journals, but don't polish it. The seal surface needs some *tooth* on it to carry oil to lubricate the seal. A seal which runs against a dry surface will fail in short order.

If the seal surface is deeply grooved or unfit for sealing for some reason or another, you'll have to replace the damper, or better yet, install a sleeve over the original seal surface. This sleeve is a relatively inexpensive part available at most automotive parts stores. The sleeve is approximately 0.030-in. thick and the lip-type oil seal can comply with this additional thickness, so there's no adjustment necessary to the seal when using a sleeve. To install the sleeve, clean the damper's seal surface and remove any burrs. Coat the original seal surface with silicone sealer to prevent oil leaking between the sleeve and the damper. Support the damper so the rubber bond will not be loaded and drive or press the sleeve on. Place a flat plate—steel, aluminum or wood—across the end of the sleeve as it sits on the end of the damper, to spread the load. Make sure the sleeve is started straight and install it until it's flush with the end of the damper shaft. Wipe any excess sealer off and the damper is as good as new.

Install the Front Cover and Damper—Before covering the front of your engine with the front cover, lubricate the fuel-pump cam with some moly. Spread it evenly around the cam's friction surface. Wipe off any oil that's on the engine block front-cover mounting surface. Use some thinner to get it really clean so the gasket sealer will seal. In the event that your engine may be torn down for another rebuild after thousands of miles of dependable use, prepare the damper so it can be easily pulled. Do this by applying anti-seize compound to its bore. It will not only help in the removal of the damper, it acts as a lubricant during the installation preventing either the damper or the crankshaft extension from galling.

Now for the cover. Apply sealer to the cover's back-side and lay its gasket in place, then apply sealer to it. Locate two of the mounting bolts that don't double as water-pump bolts—the shorter ones. The two bolts used for mounting the front cover are 5/16-18, 1- and 2-in. long. They are both UBS-style (uniform bearing strength) bolts. They have washers built into the head of the bolt. Fit the cover loosely to the engine using the two bolts after you've applied sealer to the bolt threads. Provide the lubrication for the seal by applying oil to the damper's sealing surface, then install it onto the crankshaft. Rotate the damper until it engages the key. Adjust the cover so it lines up with the damper and slide it on as far as you can by hand. The cover will be properly positioned now, so you can snug it down and install the other bolts, with exception of the ones which also secure the water pump. Remember, if your cover uses a sheet-metal timing pointer, there'll be a hole behind it for a cover bolt—don't miss it. Torque the bolts 12—15 ft. lbs.

You can now install the damper bolt and washer for good. It'll take a few turns to draw the damper down against the front of the oil-slinger and the timing sprocket. When the bolt snugs down, torque it 70—90 ft. lbs. You'll find out that you can't begin to apply this amount of torque without turning the crankshaft, so use one of your large screwdrivers to keep it from turning by locating the end of the screwdriver between the bottom of the block and the front crankshaft counterweight.

OIL PUMP AND PAN

Now's the time to seal up the bottom end of your engine. If it's not already there roll your engine over on its back. Before installing the pan, you'll have to install the oil pump first—it's just a tad difficult afterwards.

Install a New Driveshaft—If you haven't already done so, get a new oil-pump driveshaft. *Don't even think about using your old one unless it has had only a few thousand miles of service.* Just keep it around long enough to compare it to the new one to confirm its length and the location of the little Tinnerman retainer—the stamped metal ring which is pushed down over the shaft. It performs a very important locat-

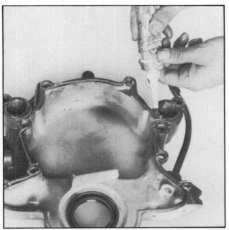  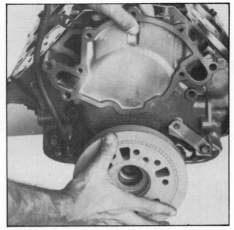

Apply sealer to your front cover, then install the gasket and still more sealer. Do a good job here because it's sealing both oil and coolant. Using two bolts, *loosely* install the cover to the engine. To center the cover, install the damper pulley on the crankshaft after oiling its seal surface. Then tighten the cover bolts.

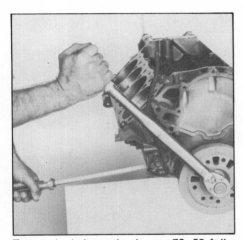

Torque the bolt on the damper 70—80 ft.lbs. Keep the crankshaft from turning with a screwdriver between the bottom of the block and one of the crankshaft counterweights.

What you'll need to install your oil pump: *new* driveshaft and two new gaskets. Large gasket is for pump base and other one is for pickup.

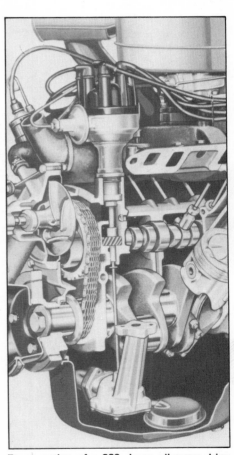

Front section of a 289 shows oil-pump driveshaft relationship to the distributor and oil pump. Note driveshaft washer at the *distributor-end*. It prevents the shaft from pulling out of the pump and falling into the oil pan when the distributor is removed.

ing function. In the installed position, it is right under the distributor-shaft hole in the block, and it prevents the shaft from being lifted out of the top of the oil pump. The importance of this feature can be realized only by those unlucky individuals who have removed their distributor only to hear an ominous clunk in their oil pan—the oil-pump drive shaft. Even worse than hearing the clunk and not discovering what it was are those who have reinstalled their distributor without the driveshaft in place and have found there is no oil pressure—hopefully before the engine is damaged.

Install the Oil Pump—You'll find 2 oil-pump gaskets, one is a little larger than the other. The big one fits between the engine block and the pump and the small one between the pump and its pickup tube. *Don't use sealer on these gaskets.* With the right gasket in place on the block, install the oil pump while making sure the pump's driveshaft engages the pump. Install the 2 3/8-16 x 1-1/4-in. long bolts with their lock washers and torque the bolts 23—28 ft. lbs. Double check to make sure the driveshaft can't be pulled out the top of the pump. If it can, slide the little retaining washer up the shaft in relation to the engine until it's just short of rubbing on the block when the shaft is firmly seated in the pump. If the pump's pickup tube is not installed on the pump, install it now using the remaining gasket—again, no sealer. Use two 5/16-18 x 7/8-in. long bolts with lockwashers and torque them 10—15 ft. lbs.

Install the Oil Pan—With the oil pump in place, take one last look at the bottom end of your engine and proceed to install its oil pan. Gather up all the parts you'll need to do the job first. You'll need the pan, the front and rear pan seals, the right and left cork gaskets and the attaching bolts. You'll need 18 1/4-20 x 3/4-in. long bolts and 4 5/16-18 x 1-in. long bolts, all with lock washers. The 4 large bolts are required to pull the pan down against the resisting force of the rear-main-bearing-cap seal and the front-cover seal—the force required to do it would strip a 1/4-20 thread.

With the block upside down, apply the adhesive-type sealer to the block's gasket surface, then fit the cork gaskets to the block—after you've figured out which one, and in which direction, goes where. One

Position driveshaft so its washer and pointed end are up in relation to the engine, and the end is inserted in the block's distributor-shaft bore. Lay oil-pump gasket on the block and install pump. Make sure driveshaft end is inserted into pump's drive socket. Don't forget the lockwashers. Torque mounting bolts 23—28 ft.lbs. Install the pickup with its gasket, bolts and lockwashers to complete the pump installation. Torque pickup bolts 10—15 ft.lbs.

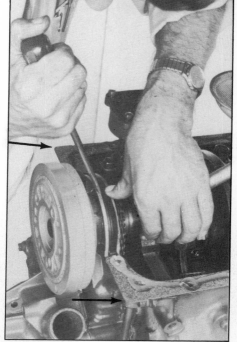

Right and left pan gaskets are different. Sharp corners go to the front (arrow). Fit the ends of the front and rear seals to the cork gaskets and force them into their grooves with a large screwdriver and a lot of thumb pressure. Use sealer only on the pan side of the gaskets.

tip, the sharp outside corners go to the front of the engine. After the gaskets are in place, fit the front-cover and rear-main-cap seals into place. These rascals can be tough. After you've figured out the long one goes in front, lay one—any one—in its groove and fit its ends to the cork gasket. The seal overlaps the gasket to make a good seal at the junction of the seal and the gasket. One projection, or tongue goes between the gasket and its groove and the other projection overlaps the gasket. With the seal in place, work it down into its groove with your thumb and a screwdriver. Use the screwdriver to force it into place and your thumb to hold it there. Start at one end of the seal and work around to the other end. After they are both in place and you've applied sealer to the cork gasket only, you can install the pan.

Before setting the pan on the engine, take this opportunity to soak the timing chain with oil. This'll give it some lubrication when your engine is being cranked for its initial firing. Now you can install the oil pan, but be careful not to move the gaskets. Loosely install the pan bolts and *tighten them progressively*. This is a good application for a speed handle. Don't overtighten the bolts, you'll distort the oil-pan flange. You'll notice that in going around tightening all the bolts that the cork gasket relaxes after a short while. Go around again until they begin to feel firm, then they are ready to be torqued. Check the front and rear seals to make sure they were pulled down into their grooves, then torque the pan bolts. Torque the 18 1/4-20 bolts 7—9 ft. lbs. and the 4 5/16-18 bolts 9—11 ft. lbs. Because of the way the cork gaskets relax, go around the pan several times, until the bolts don't turn anymore at their specified torque. Don't distort the pan by over-tightening the bolts.

Now that the crankcase is sealed off by the oil pan, close off the distributor- and fuel-pump holes so a bolt, nut, wrench or anything of the sort won't inadvertently end up in the oil pan. A rag stuffed in the holes does this very effectively—or some duct tape. Also, don't forget the oil-pan drain plug. Install it after you've checked its seal to make sure it's in good shape. It's always embarassing to be filling your newly rebuilt engine's crankcase with nice clean oil just to have it pour out the bottom of your engine at the same time.

INSTALL YOUR WATER PUMP

Now's a good time to install your water pump because it makes a nice handle for wrestling your engine around. You'll find 2 gaskets in your gasket set. Depending on whether you have the aluminum- or cast-iron pump, the first gasket fits between the spacer and the pump or the cover-plate and the pump, respectively. The aluminum pump isn't sealed from the front engine cover, its spacer, or plate has a large clearance hole for the impeller whereas the cover plate for the later cast-iron pumps completely seals the back of the pump from the front cover.

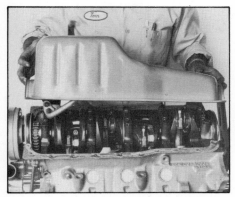
Carefully position oil pan on block and install all oil-pan bolts before tightening any of them. Snug the pan bolts down several times because the cork gasket relaxes. Torque the 1/4-20 bolts 7—9 ft.lbs. Larger 5/16-15 bolts adjacent to the front and rear-main bearing caps: 9—11 ft.lbs.

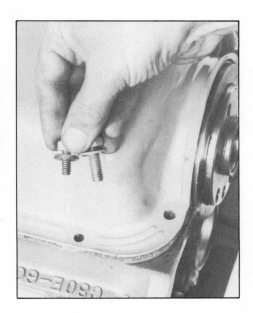

If you haven't already done so install the rear cover to the pump if yours is the cast-iron one. You'll need one of the gaskets between the pump and the cover, the one that seals the pump-outlet holes on the outboard side only. Use the adhesive-type sealer here on both sides of the gasket and secure the cover to the pump with the 2 5/16-18 x 1/2-in. long special bolts. Their hex size is 5/8 in. which is low profile at only 3/32-in. high so they clear the front cover. With the cover secured in place and the bolts torqued to 15 ft. lbs., install the pump assembly to the cover using the remaining water-pump gasket and the adhesive sealer. Install the water-pump and torque its bolts 12—15 ft. lbs. Remember to put sealer on the threads of the bolts which thread into the front face of the engine block— the bolts which double as front-cover bolts.

For the earlier aluminum pump, use the adhesive sealer to fit the pump-to-spacer gasket to the pump, then the spacer to the gasket, the spacer-to-front cover gasket to the spacer and finally the whole assembly to the engine. Use two of the

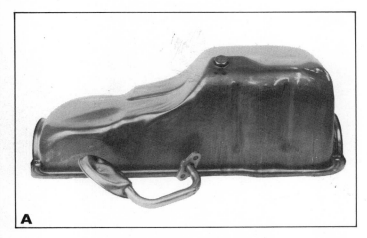
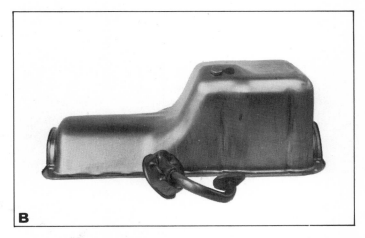

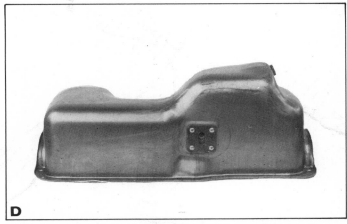

Four of the many oil pans and their pickups that have been installed on the small-block Fords: A. Fairlane/Comet, Mustang/Cougar, Maverick, Granada/Monarch, Versailles and full-size Ford and Mercury. B. Fairmont/Zephyr. C. Econoline. D. 4X4 pickup and Bronco. Make sure your oil pan fits the installation if you're doing an engine swap.

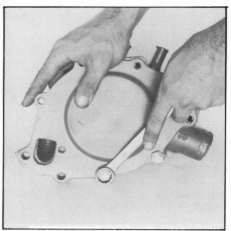 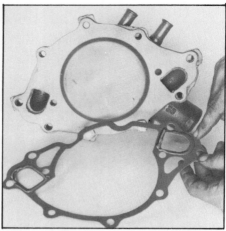

Install the cover plate on a cast-iron water pump. A gasket with sealer on both sides is required between the cover and the pump housing. Use low-profile bolts to retain the plate to the pump. Gasket between the pump assembly and the front cover needs sealer on both sides around the water-passage openings only. For pumps not using a cover plate, the gasket needs a full bead of sealer front and back. Use sealer on the mounting bolts and torque them 12—15 ft. lbs. Some of the bolts will have to come out later for installing your accessories.

mounting bolts to line up the 2 gaskets and the spacer to the pump. Use sealer on the long bolts which double as front-cover mounting bolts and torque them all 12—15 ft. lbs. Be prepared later on to remove some of the water-pump bolts because they'll not only serve to retain the water pump and the front cover, they'll also mount some of the engine accessories, thus they will be doing triple duty.

CYLINDER HEADS

It's now time to begin buttoning up the top side of your engine, so position it upright. It'll be unstable, so you should block it up under the oil-pan flanges so the engine sits level and won't roll over. If you don't it will get worse because of the additional weight you'll be adding to its top in the form of the cylinder heads. If you are using an engine stand, you've got it made in this department.

Coat your head gaskets with high-temperature aluminum paint. Install them right-side up and in the right direction over the dowel pins as indicated on the gaskets (arrow).

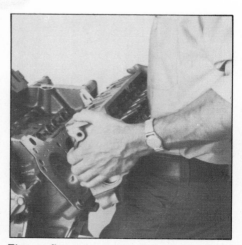 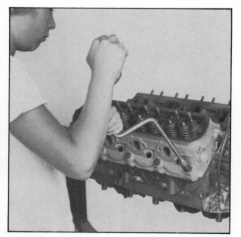 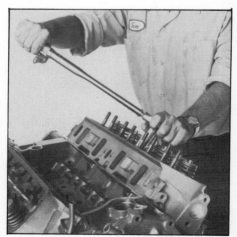

Time to flex your muscles. Locate the cylinder heads on the dowel pins. Oil the head bolts. I turned the speed handle over to my son Jeff for this job. Torque the head bolts in the proper steps and sequence.

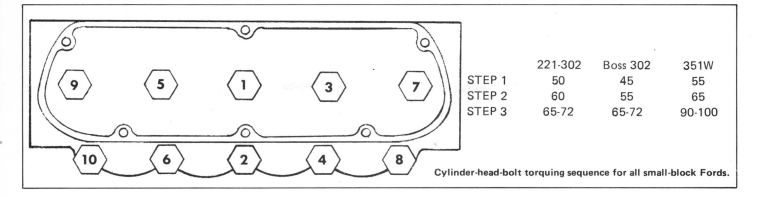

	221-302	Boss 302	351W
STEP 1	50	45	55
STEP 2	60	55	65
STEP 3	65-72	65-72	90-100

Cylinder-head-bolt torquing sequence for all small-block Fords.

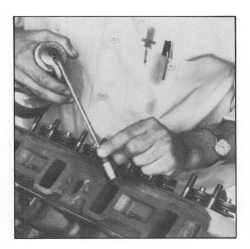

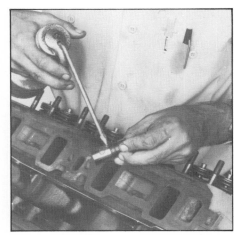

Liberally oil your lifters before installing them. If they are the hydraulic type, prime them. When installing used lifters with their original camshaft install them in their original bores.

Get your cylinder heads out of storage, the 2 head gaskets from your gasket set and the head bolts—20 grade-8 bolts to be exact. The 351W uses 1/2-13 head bolts: 10 are 2-1/2-in. long and the other 10 are 4-in. long. All other small blocks use 20 7/16-14 bolts, 10 2-9/32-in and 10 3-29/32-in long.

Clean the Gasket Surface—To ensure the best possible head-gasket seal, clean the cylinder-head and cylinder-block gasket surfaces with solvent to remove any oil or grease deposits. Check to make sure all 4 hollow cylinder-head-locating dowel-pins are in place. They are a must to locate the cylinder heads accurately during installation, so if they are missing replace them. Do this by inserting the straight end in the block with the chamfered, or tapered end projecting outward and tap it into place. You'll feel the pin bottom in its hole when it's fully installed.

Paint the Head Gaskets—One of the best and easiest to apply head-gasket sealers is high-temperature aluminum paint applied with a spray can. Evenly coat both sides of your head gaskets with it. After the paint has dried, position the gaskets on the engine block by locating them on the dowel pins with the *FRONT designation on the gaskets to the front and up.* If you get them mislocated, you'll be wondering why your engine is overheating. Misrouting of coolant from the block to the cylinder heads due to water-passage blockage at the wrong end of the heads an over-heating engine.

Install the Heads—Flex your muscles now and lift the cylinder heads into position on the block. Locate them over the dowel pins and make sure each is down against its head gasket. Install the head bolts after coating their threads with a sealer—five short ones along the bottom of each head and five long ones in the rocker-arm area. When torquing the head bolts, they must not only be done in sequence, they must be done in three stages or steps as described in the above sketch and table. Remember to do each stage of torque in the proper sequence, not just the final one. Refer to the sketch for the right sequence. To keep the cylinders free of moisture and dirt, install an old set of spark plugs in the heads, some you won't care whether they get broken or not. Secure the dip-stick tube to the front of the right-hand cylinder head with its bolt and washer.

VALVE TRAIN

To complete your engine's valve-train assembly after the installation of the cylinder heads, you'll need 16 valve lifters, pushrods, rocker arms, fulcrums and nuts. You'll also need your squirt can full of oil and some oil additive.

Install the Valve Lifters—Start putting the final touches on your valve train by installing the valve lifters. If you have new ones, they can be installed in any order, otherwise they'll have to be installed in the bores they were removed from. Prime your hydraulic lifters. Do this by filling them with oil using your squirt can. Force the oil in the side of each lifter with your squirt can until it comes out the top hole. After priming a lifter, oil its OD, coat its foot with moly-disulfide and install the lifter in its bore.

The Pushrods Are Next—With the lifters in place, slide the pushrods down through their holes in the heads into the center of their lifters. If you are installing new pushrods they can go in any direction. If you are installing your old ones, *install them end-for-end*—the end that operated at the lifter should now be at the rocker arm. You can tell which end was which by looking at the wear pattern on the ends of the pushrods. The area worn on the lifter end will be much less than that at the rocker-arm end.

Install the Rocker Arms—Your engine will be equipped with one to two types of rocker arms if you have a 1978 or later 351W, the cast rockers with spherical-ball pivots or the later style stamped-steel ones with the integral rocker-pedestals

Ford's oil conditioner is just one of the many oil supplements that can be used when assembling your engine.

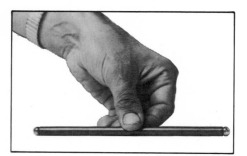

Regardless of whether your pushrods are new or not, check to see if they are straight by rolling them on a flat surface. I'm using a kitchen counter here. If more than 0.020-in. feeler gauge fits between the surface and the center of the pushrod, replace the pushrod. DON'T TRY TO STRAIGHTEN IT.

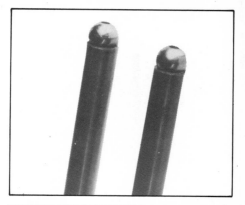

Wear pattern for rocker-arm end (top) of a pushrod extends farther around ball than for the lifter end (bottom). Swap ends when reinstalling used pushrods.

and cylindrical pivots. Otherwise, you'll have the cast rocker arms and ball pivots. Regardless of which type rocker arms your engine has, they should be prelubricated prior to their installation. Apply oil additive to their pivots and pushrod contact surfaces and moly grease to the valve-stem tips. Begin the installation by fitting the rocker arms, pivots and nuts or bolts to the head. If you are installing your old rockers and pivots, remember to keep the pivots mated with their rocker arms.

Adjust the Valves—A valve train must be adjusted when its valve is in the fully closed position. How the adjustment is done depends on the type of valve train in your engine. For example, all small blocks produced through mid-1968 had fully adjustable valves. How far the rocker-arm pivot is moved up and down its stud with the adjusting nut determines a valve's adjustment. All 302s produced after mid-1968 and all 351Ws do not have fully-adjustable valves. Their valves are adjusted by the installation of different length pushrods—0.060-inch longer or shorter than the standard length.

Another factor in determing how a valve is adjusted is the type lifters used—mechanical or hydraulic. Mechanical lifters were used in the HP289 and Boss 302. All others used hydraulic lifters.

Valves operated by mechanical lifters are adjusted so there is a specified clearance, or *lash* between the rocker-arm and the tip of the valve when the valve is closed to ensure full valve closure. Hydraulic lifters are designed to operate without this lash for quieter engine operation and lower maintenance requirements. They also require less precise valve adjustment than the mechanical type.

Index the Crank—Regardless of which type of valve-train your engine has, each valve must be adjusted when it is in the fully closed position—its lifter must be on the base-circle of the camshaft lobe. Both valves of a cylinder are closed when it is at the TDC of its firing position, however at least one of the valves will be fully closed during a cylinder's other three cycles. So, rather than having to index a crank eight times to the TDC of each cylinder, the valve closings have been combined so it needs to be indexed only three times. Because of the difference in firing orders between the 351W and the other small blocks, the valves adjusted in the three positions will be different. Start by positioning the crank on number-one cylinder TDC.

Fully Adjustable Hydraulic—Finish installing and adjusting a valve by running the adjusting nut down its stud until the slack is taken up in the rocker arm and pushrod. Wiggle the rocker arm to make sure it's in its freest position. When the slack is taken up, give the adjusting-nut another 3/4 turn. The valve is now adjusted and you can go on to the next one.

Fully Adjustable Mechanical—Adjusting mechanical lifters as opposed to the hydraulic type is a whole different deal. First there is lash, the clearance between the valve tip and the rocker arm. Lash must be 0.018 in. when the engine is at its operating temperature, but it's impossible to check *hot lash* at this point. However, the valves must be adjusted to something so you can get on with your engine assembly and to enable you to start your engine initially. This something is *cold lash*. Adjust it to 0.022 in. by using your feeler gauge between the valve tip and the rocker arm. Wiggle the rocker arm to be sure all the slack is out of the valve train, but so there is a slight amount of drag on your feeler gauge as you slide it between the valve and rocker arm. With the cold lash set, you'll be able to run your engine so final hot lash adjustments can be made.

Positive-Stop Hydraulic—The only method which can be used to adjust the valve train using positive stop-type studs or the later type with stamped rocker arms is to install longer or shorter pushrods. This should not be required unless one of your valve seats required extensive grinding to clean it up, thus causing the valve stem to extend out of its guide more. This means a shorter pushrod may be required. The same applies if your heads were milled, otherwise you shouldn't have any problems. Regardless of what should be, check your valve-train settings in their installed positions to be sure.

There are two methods for checking positive-stop type valve trains. The first one requires a special tool, one that will be difficult to find. It rotates the rocker arm against the pushrod while the valve is in its closed position by hooking under

A new rocker arm, pivot and nut. Lubricate your rocker arms well as you install them. Dipping a clean screwdriver in oil is an easy way of doing this.

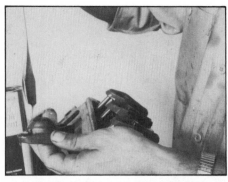

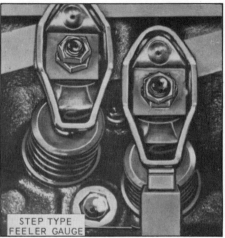

Boss 302 Valve adjustment with a step-type feeler gauge. The Boss 302 and HP289 valves have to be static adjusted now during assembly, then later when the engine can be run for hot lashing. Photo courtesy Ford.

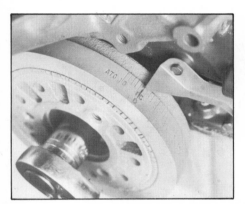

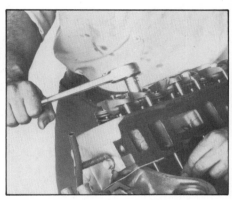

Start with cylinder 1 on TDC when adjusting your valves. When adjusting hydraulic valves, rotate the pushrod to feel for takeup, then turn the nut another 3/4 turn.

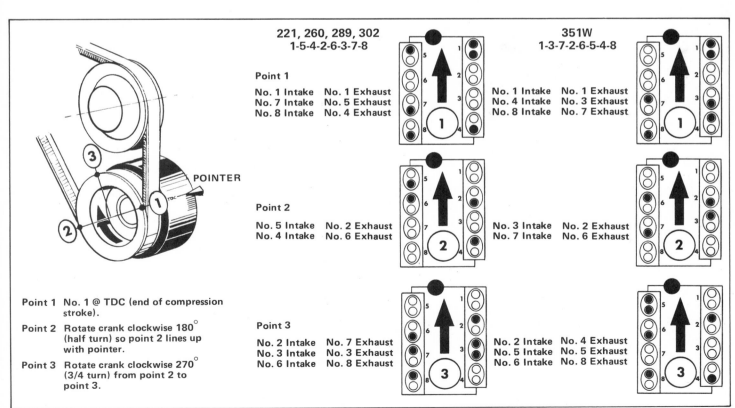

Point 1 No. 1 @ TDC (end of compression stroke).
Point 2 Rotate crank clockwise 180° (half turn) so point 2 lines up with pointer.
Point 3 Rotate crank clockwise 270° (3/4 turn) from point 2 to point 3.

Use this chart to save time when adjusting your valves. Adjust the valves indicated in the positions shown. If it confuses you, go through your engine's firing order to do the adjusting.

Intake-manifold studs on the 302 are handy for locating head-to-manifold gaskets. If your engine isn't equipped with these studs you can order them under Ford part number 382198-S. You'll also need the 374031-S nuts. Stud size is 5/16-18 thread x 2-3/8" long.

Where 90% of all small-block Ford oil leaks come from. Use adhesive sealer here with a dab of silicone sealer where the side and end gaskets join. Don't forget sealer around the water passages.

the rocker arm on the valve-side of its pivot and pushing down on the rocker arm on the pushrod-side. The purpose of this is to collapse the lifter while the valve-tip-to-rocker-arm clearance is checked with a feeler gauge—the gauge can't be more than 3/8-in. wide when used with rail-type rocker arms. The tool you'll need for rotating the rocker arm is Ford's special tool 6513-AC. 302 engines should have 0.090—0.190-in. clearance whereas the clearance for 351Ws is 0.106—0.206-in. Like I said, you'll have a difficult time finding this tool so let's take a look at the second alternative.

Just as rocker-arm-to-valve-stem clearance is directly proportional to lifter collapse, so is the vertical movement of the rocker-arm nut. How many times this nut is turned collapses its lifter a given amount. This basic fact eliminates the need for any special tool, thus making the job much easier and less involved.

To determine if you need different length pushrods count the number of rocker-arm-nut turns it takes to *bottom* each rocker-arm pivot on its stud shoulder from the point all slack is taken out of the pushrod and rocker arm, but before any lifter plunger travel occurs. Wiggle the rocker arm and rotate the pushrod to check for slack, and look at the lifter to see if its plunger has moved. Run the nut down while keeping track of the turns. When the pivot bottoms, or turning torque increases, stop turning and counting. If you're in the tolerance range finish the job by torquing the nut 17—23 ft. lbs. Different tolerances are called for 302 and 351 engine, consequently there will be a different number of turns for the two engines. It should take 0.8 to 1.75 turns for 302 engines and 1 to 1.8 turns for the 351W. If fewer turns are required, you need a shorter

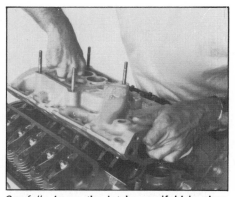

Carefully lower the intake manifold in place. DON'T SHIFT THE GASKETS OUT OF PLACE.

pushrod and a longer one if more are required.

INTAKE MANIFOLD

Before covering the lifter valley up with the intake manifold, pour the rest of your oil additive over the valve lifters and on the rocker-arm pivots. You won't need it any more and you might as well put it where it'll do the most good. You'll need 4 gaskets to install the intake manifold, 2 asbestos gaskets to seal the manifold to each cylinder head and 2 cork gaskets to seal both ends of the manifold to the block. These are the ones that classically leak, so pay particular attention to them, otherwise it's very likely you'll have a healthy oil leak at one or both ends of the manifold to the block. In addition to the gaskets, you'll need two types of sealer, the adhesive type and silicone. Finally, you'll need the attaching hardware: 12 grade-8 5/16-18 x 2-1/4-in.-long bolts or 8 5/16-18 x 2-1/4-in.-long studs with their nuts if your en-

All but the end bolts use washers under their heads for sealing—don't forget them. Torque the manifold bolts in the proper sequence.

gine is a 302. 351Ws use 12 bolts and 4 studs, all with 5/16-18 threads.

Put the Gaskets in Place—Before the intake manifold can be installed, the gaskets must be located on the heads and the block. If you have a 302, make sure the 4 studs are threaded into the head. Locate each one in the second hole in from both ends of each head. These are great for positioning the manifold-to-head gaskets and for guiding the manifold in place when you're installing it—too bad the other engines aren't equipped with the studs. Run a bead of sealer around the water passage openings at the end of both heads and lay the manifold-to-head gaskets in place. They are designed so both ends of their lower corners hook into projections on the cylinder-head gasket.

Now comes the "problem child," the manifold-to-block gaskets. These gaskets cause 90% of the small-block Ford oil leaks. The gaskets are installed neatly on the block, then the manifold is carefully laid in place on the engine. Everything looks good up to this point. As the manifold is tightened down, the gaskets begin to squeeze out from behind the manifold and the block to the inside where they can't be seen. The eventual result is an oil leak.

To install your manifold-to-block gaskets successfully you'll have to adhere them to the block so they *can't* squeeze out. This is where gasket adhesive or weatherstrip adhesive is invaluable. Before applying any sealer to the block, clean the gasket surfaces with lacquer thinner, then you can apply a thin bead of sealer to the block's gasket surfaces and spread it out evenly with a finger tip. Do the same thing to the mating side of the gaskets so when the two surfaces contact each other they stick. However, before installing the gaskets let the sealer dry to the point of being *tacky*, then install them. Now apply silicone sealer around the cylinder-head-to-manifold water passages and install the gaskets on the heads. Make sure the ends of the gaskets interlock with the block-to-manifold gaskets. Some more silicone sealer around the four water-passage openings and some at the junction of the four gaskets followed by a bead of gasket adhesive sealer on top of the block gaskets and you can install the manifold. Letting the adhesive sealer get tacky before you install the manifold will further ensure the end gaskets won't squeeze out.

Install your intake manifold by lowering it carefully over the gaskets. If you have a 302, the 4 studs solve this problem, otherwise avoid shifting the manifold on the engine once it's down on the gaskets—you may slide them out of place. This is difficult to do because of the weight of the manifold, but it's also necessary. Try to sight down on the manifold and the top of the engine so you can see what you're doing. Thread the bolts into place and begin tightening them. Keep an eye on the gaskets to make sure they stay in place. Tighten the bolts and final torque them in the sequence shown in the sketch. There is a unique sequence for the 351Ws because of their four additional attachments. In sequence, torque the bolts and stud nuts in two stages, the first one 15–17 ft. lbs. and the final one 20–22 ft. lbs. Because the gaskets relax, go back over the bolts and retorque them until they've fully compressed.

Sealer Instead of Gaskets—Another method of sealing the intake manifold to the engine block is to substitute silicone sealer in place of the two cork end gaskets. This method avoids the possibility of the gaskets squeezing out, however, the *block and manifold gasket surfaces must be clean* for the silicone to work. Therefore, take special care to clean these surfaces using lacquer thinner. Now, with the

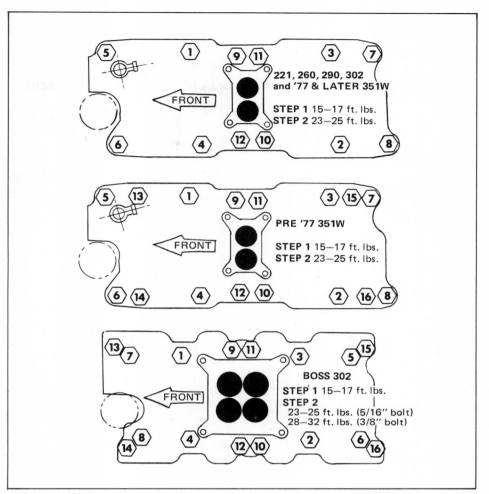

Intake-manifold torquing sequences.

cylinder-head-to-manifold gaskets in place apply a 1/4-in. bead of silicone sealer to the cylinder-block gasket surfaces—just enough so the intake manifold will contact and squeeze the bead wide enough to seal after the manifold is torqued down. Install the manifold as previously outlined being careful not to disturb the 2 cylinder-head gaskets and silicone beads, then torque the manifold bolts in sequence.

Valve Covers—Without using any gasket sealer, install the valve-cover gaskets in their valve covers. Insert the tabs on the periphery of each gasket in the notches provided for them in the valve-cover flanges. Someone was using his grey matter when he thought these up. You don't have to worry about the gasket shifting or falling out of place when installing the covers.

With the gaskets in place, install the covers on the cylinder heads. Secure them with 6 1/4-20 x 5/8-inch-long bolts with their lock washers. Don't over-torque these bolts or you'll succeed in bending the valve-cover flanges and not much else. Torque the bolts 3–5 ft. lbs. Retorque them after the gasket has had time to compress.

Exhaust Manifolds—In your gasket set you'll find 2 exhaust-manifold gaskets. They are either composition-type gaskets—asbestos and metal—or all-metal. If you have the composition type, mount the metal side against the manifold—it will be marked MANIFOLD SIDE. Use adhesive sealer on the gasket. Apply it to both sides of the all-metal gaskets, but just the metal side of the composition gaskets. Mount each manifold and gasket loosely with a short bolt at the front and rear of the manifold. Because an exhaust manifold can *float* around on the mounting surface of its cylinder head due to oversized mounting holes, it's possible that they will end up being positioned too low on the cylinder head. If this happens, there won't be enough room to use your spark-plug socket. With this in mind, snug the 2 bolts down while lifting up on the manifold to position the manifold high enough to be able to service your sparkplugs.

Install the other exhaust-manifold bolts, but don't do it indiscriminately. There may be 2 or 4 of the long manifold bolts which have studs protruding from their heads. If you have 2 of these, they go in the middle of the right manifold.

119

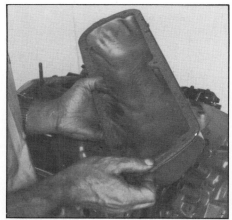 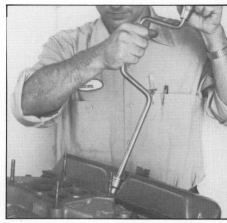

Using the valve-cover gasket *ears* to hold them in place, install the valve covers on your engine. It's not required, but if you want to use sealer, put it on the valve-covers only. To keep from bending the sheet-metal or cast-aluminum covers don't over-tighten the bolts (3–5 ft.lbs.).

Apply adhesive sealer to the exhaust manifold only when using gaskets. Use silicone sealer when you're not using gaskets. Metal side of composition gasket goes against manifold. When installing your manifold, check spark-plug socket clearance, then secure it. Torque the bolts twice. Now, and then later after the gaskets relax.

They are for attaching the heat-riser shroud. If you have 2 more they may be for attaching spark plug shrouds to the left side of the engine. I say "maybe" because some engines have these bolts, but nothing attaches to them. Anyway, one is used at both ends of the left exhaust manifold regardless of whether anything is attached or not as they come from the factory. Torque the manifold bolts 18–24 ft. lbs. for the pre-'77 351W and 13–18 ft. lbs. for the rest. Retorque them later if you have the composition gaskets. You'll find that you can give each bolt another quarter turn or so.

Miscellaneous Hardware—The major portion of the components which make up your final engine assembly are now in one piece, however there are a lot of little parts yet to be installed on your engine. There are even some that are not quite so little. What additional parts you install now must be weighed against the pros and cons. For example, how hard is it to install now as opposed to after your engine is installed, whether or not the part will interfere with installing the engine, the possibility of the part being damaged during the installation and the weight added to the engine—will your lifting device handle the additional weight? You'll have to decide what to install and what not to now. I'll describe what worked best for me and let you take it from there.

Install the Thermostat and Its Housing—Install a new thermostat in your intake-manifold coolant outlet. Make sure you have it pointed in the right direction. Which direction goes toward the engine will be indicated on the thermostat. Position the thermostat in its housing with the end indicating TOWARD ENGINE projecting out of the housing. There will be a relief for the housing flange to sit in. Apply adhesive-type sealer to the housing gasket surface—make sure you circle all the holes with the sealer— then install the gasket on the housing. Coat the opposite side of the gasket with sealer and install the thermostat and housing assembly to the intake manifold using two 5/16-18 x 1-inch-long bolts with lock washers. Torque the bolts 12–15 ft.lbs.

If you removed the vacuum control valve from the thermostat housing, re-install it. Coat the thread with sealer first, then install it snugly in the housing. Position the valve so it points 90° to the right.

Intake-Manifold Hardware—If your intake manifold was stripped during the reconditioning process, reinstall the water-temperature sending unit, the heater-hose inlet fitting and the vacuum manifold if your engine was so equipped. Also, if you have the 221 or 260 engine with PCV system, install the crankcase outlet fitting at the back of the manifold. All but the PCV outlet use pipe threads, consequently they don't have to be torqued to seal, just tighten them. If they have to be installed in any specific direction, use sealer on their threads. If you have Teflon-tape sealer, use it.

When installing the heater-hose inlet fitting, rotate it in the manifold so its hose end points 90° to the right. Don't overtighten it, it just has to be snug to seal. The reason for rotating in this manner is to facilitate better heater-hose routing.

As you can tell by the many parts left

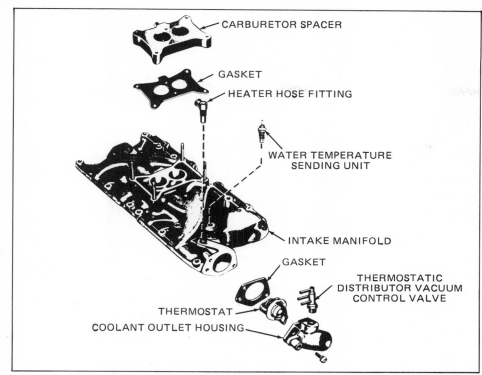

Some of the hardware attached to an intake manifold. Drawing courtesy Ford.

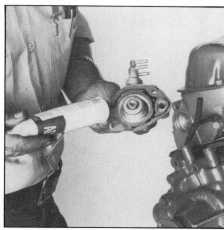

Thermostat goes in the housing before the gasket goes on. Make sure you have the thermostat pointed in the right direction.

to install on your engine, there are quite a few that can be installed now rather than after it's back in its engine compartment. However, let's get on with the actual engine installation because most of the remaining parts and components are easier to install just prior to the time your engine is lowered into its compartment while it is hanging in mid air—or after it's in place.

If your manifold-to-heater-hose fitting looks anything like the one at left, replace it! It's another example of what using water only as coolant does to cooling-system components.

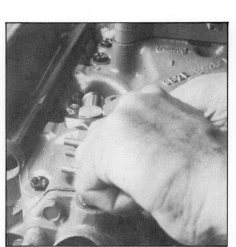

A little bead of sealer around the first couple of threads and you can install your heater-inlet fitting. Rotate the fitting to point between the carburetor and the right valve cover for proper hose routing.

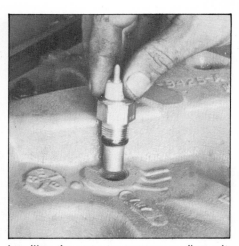

Installing the water-temperature sending unit. Seal its threads too.

8 Distributor Rebuild

1968 302-2V equipped with an advance/retard distributor. This engine has served as the base V-8 in most Ford passenger cars and light trucks at one time or another since 1968. Photo courtesy Ford.

Due to the general lack of information concerning distributor rebuilding, this chapter explains how to determine if your distributor needs to be rebuilt, then shows how to do it. Before tearing into your distributor, find out which one you have. Several types have been used on small-block Fords, however rebuilding is basically the same regardless of the type of distributor.

Two *basic* types of distributors are installed in small-block Fords: *breaker-point* type and *breakerless* (solid-state) type. The difference is how they produce a spark. The breaker-point distributor works by closing and opening a set of breaker, or contact points on a breaker plate. The breakerless distributor accomplishes the same thing electronically through a magnetic triggering device and an amplifying module.

Otherwise the two are basically the same. Both are driven at 1/2 crankshaft speed by a gear on the camshaft through a shaft which runs in bushing/s in the distributor housing. Distributors installed in engines prior to 1970 may or may not use two replaceable bushings. Beginning in 1970 all distributors use the top bushing only and depend on the nose of the bottom end of the distributor shaft running in the block to support the gear end of the shaft.

On the upper end of the distributor shaft is a cam which operates breaker points, or an armature that triggers the magnetic pickup. The cam or armature is free to rotate on the shaft within limits. This rotation is controlled by centrifugal advance weights which cause spark timing to change with RPM.

Mounted on the same shaft above the cam is the rotor, common to both types of distributors. As it rotates, it completes the circuit between the coil and each spark plug in the correct firing order. The distributor base plate mounts the breaker plate or the magnetic-pickup assembly. Otherwise the two types of distributors are the same.

Small-block distributors are the *dual-advance* type. To change spark timing, they have centrifugal-advance weights to rotate the cam or armature and one or two *vacuum diaphragm/s* which rotate the breaker plate or magnetic pickup assembly around the center shaft. One exception is the *dual-point* distributor used on all HP289s and Boss 302s. They use two sets of breaker points instead of the usual single set. The HP289 does not have a vacuum-advance diaphragm. Its breaker plate is fixed.

Another variation occurs at the vacuum-advance diaphragm in the dual-advance distributors. All pre-1968 distributors use a *single diaphragm* operated by carburetor venturi vacuum. Beginning in 1968, the *dual diaphragm* was used for emission-control reasons on engines sold in some areas. This unit uses the carburetor-vacuum-actuated diaphragm. It also has a vacuum-retard diaphragm in tandem to retard the distributor when there is high manifold vacuum—at idle and during deceleration when the throttle is closed. Other than the diaphragm, the two distributors are mechanically the same even though they have different advance/retard curves. All solid-state distributors use the dual diaphragm exclusively. Those in 1977 models and later years are equipped with a larger diameter distributor cap with an adapter between the distributor housing and the cap. These also use a unique rotor, and wiring harness leads with spark plug-type connectors at both ends.

Does It Need Rebuilding?—As with your engine, you must first determine if your distributor *needs* rebuilding. Remove the cap and rotor, then wiggle the upper end of the shaft back and forth—or at least try to. If it moves a noticeable amount, something is worn too much. It could be the ID of the cam or sleeve, or the upper shaft bushing. To isolate the two, remove the cam from the end of the shaft using the procedure on page 125, then try wiggling the shaft. If it still moves you know it's the bushing. If it doesn't, then it's the cam or sleeve ID. The problem could be a worn shaft, but it'll probably be the bushing or the cam or sleeve. To make sure it's not the shaft, mike its upper portion that runs in the bushing and compare this figure to what you get when miking an unworn section of the shaft. To do this you'll have to remove the shaft, so refer to the teardown section later in this chapter. First I'll describe a more accurate way of determining if your distributor needs rebuilding.

Let Performance Be the Judge—Let your distributor's performance be the determining factor as to whether it needs attention or not. To do this, take it to a shop which has a distributor test set. Check for proper centrifugal advance and vacuum advance/retard—applies to both types of distributors—and dwell variation for the breaker-point type. Chances are your centrifugal-advance mechanism will check out OK. If it doesn't, a minor adjustment should correct the problem. Possibly the weight pivots need to be cleaned and

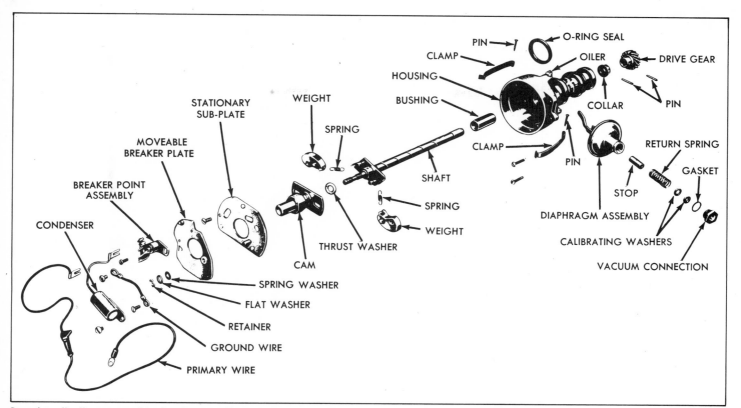

Complete distributor assembly showing centrifugal- and vacuum-advance components. Centrifugal weights are the pre-1965 style which operate the cam with pins protruding from each weight into slots in the cam plate. Photo courtesy Ford.

Checking a distributor out the right way—on a distributor test stand. Centrifugal advance is being adjusted by bending the advance-stop pin.

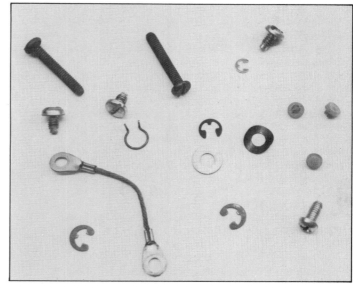

Some small parts used in most distributor assemblies. Be careful when disassembling yours so you don't lose any of them. Hours can be spent finding a replacement.

lubricated if they are the bushingless type. As for the vacuum advance/retard mechanism, you have more of a chance of encountering trouble. This usually occurs in the form of a leaky vacuum diaphragm, meaning it will have to be replaced. With a breaker-point distributor, another possibility is the movable breaker-plate may be hanging up. This will either be caused by a bent breaker-plate, or the three nylon buttons on which the plate slides are worn. You may be able to straighten the plate, but worn buttons mean breaker-plate replacement—you can't buy the buttons.

You also need to determine the condition of your distributor's upper bushing and the breaker-plate pivot. This is done by monitoring the dwell on the test set. If the dwell varies, the cam is moving in relation to the points, or the points are moving relative to the cam. When the cam and points are farthest from each other, the points stay closed longer, thus *increasing the dwell*. The reverse is true when they come closer together. The likely cause of this is a worn upper bushing, or possibly excess wear between the cam and the distributor shaft. More than likely it's the bushing. Breaker-plate pivot wear can also cause dwell variations. The pivot is a brass bushing pressed into the moveable breaker plate which pivots on a pin

123

attached to the base plate. Instead of the cam moving in this instance, the points move in relation to the cam, but the effect is the same. So, if the breaker-plate pivot is worn so there are dwell variations, the *breaker-plate assembly* should be replaced. The bushing is not serviced separately.

REBUILDING YOUR DISTRIBUTOR

Before you begin tearing down your distributor to rebuild it, make sure your Ford dealer has the bushing/s in stock. He'll prefer selling you a new or rebuilt distributor rather than the bushing/s so you can rebuild your own distributor. Buy upper bushing B8QH-12120-A and, if you need it, the lower one C5AZ-12132-A *before* you pull your distributor apart. If you encounter problems with obtaining these, you may have to reconsider and start shopping around for a replacement distributor. Assuming you can come up with all the parts you need for the rebuild, get on with the job.

Disassembling the Breaker-Point Distributor—After removing the distributor cap and rotor, remove the breaker points and condenser. Keep the small parts, nuts, screws, clips and the like in a small box, can or paper container so they don't get lost. It's a pain to find replacements. Remove the vacuum-diaphragm assembly after disconnecting its link from the breaker plate. Use a small screwdriver to pry the clip off and keep a finger on the clip so it doesn't fly away.

The breaker-plate assembly can come out in one piece, but I've shown it being removed in pieces. The fixed-plate HP289 distributor can be removed after its two attaching screws are removed. As for removing a fixed breaker plate, remove the C-clip, flat washer and spring washer from the pivot, then lift the breaker plate out. This will expose the base plate which you can now remove. Don't lose the three nylon buttons in the plate, they pop out easily.

Check the breaker-plate pivot for wear now by fitting the two plates together loosely and trying to slide the breaker plate sideways. If you can feel any movement, replace the complete assembly.

Disassembling the Breakerless Distributor—If you have a breakerless distributor, tear it down similarly. Remove the cap and rotor. On 1977 and later models remove the cap adapter.

Remove the *armature*—the wheel with eight vanes—by prying it up off the shaft using two screwdrivers. Don't lose the *roll-pin* which keys it to the shaft. Remove the vacuum-diaphragm assembly by unhooking it from the magnetic pickup and unbolting it from the housing. Now remove the magnetic-pickup assembly. Use your finger nail to remove the light wire retainer, then lift the pickup assembly off the base plate. The base plate can be removed now by undoing the single remaining screw.

With the base plate out of the way, your distributor looks nearly identical to a breaker-point type in the same state of disassembly. The difference is the straight sleeve which accepts the armature rather than the cam used for operating a set of points. The remainder of the teardown is the same for both types.

Disassembly From the Base Plate Down—You can remove the main distributor shaft at this point, but I've shown removing the centrifugal-advance weight and springs first. However, before you remove the springs, weights or cam or sleeve, mark them so you'll have a reference for reinstalling them in their original positions. A dab of light-colored paint on *one spring and its two attachment points* and *one of the weights and the pin it is fitted to* will do the job. As for the cam or sleeve, it

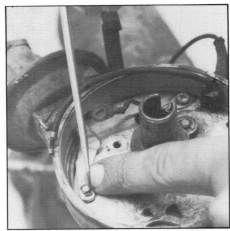 

Keep a finger on the C-clip when removing it. If one "flys" off you can just about forget finding it. Lift the diaphragm link off its post, then slide the diaphragm assembly off the distributor housing.

Remove the primary wire *in* through the distributor housing. Replace it if you find the insulation damaged.

 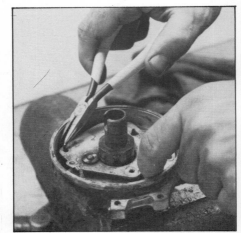

Removing the C-clip, flat washer and spring washer from the breaker-plate post. This frees the breaker plate so it can be lifted off.

Base-plate screw also retains breaker-plate ground wire. Don't lose the nylon wear buttons which plug into the base plate. If you do or if they are worn badly, you'll have to replace your distributor or find one you can rob parts from.

Prying the armature off a solid-state distributor shaft. Two screwdrivers and some light pressure should do it.

Magnetic-pickup retaining clip is best removed with your fingernail. Remove vacuum-advance diaphragm and lift out the pickup.

has two different-length notches at the edge of its plate for controlling maximum centrifugal advance. The notch with the correct advance is located over the tab which is fitted with a Hypalon® sleeve. Put a dab of paint at the edge of this slot. Now, all you have to do is be careful not to remove the paint while you're cleaning the parts.

1965 and Later Cam, Spring and Weight Removal—When unhooking the centrifugal-advance return spring, stretch it *only* far enough to lift if off the pin. The cam or sleeve can be removed after you have the springs off. Remove the retainer from the center of the cam or sleeve. Lift the lubricating wick out. Use small needle-nose pliers to lift one end of the spring out of its groove and a screwdriver to keep it from slipping back in while you lift the other end out. After removing the retainer, lift the cam or sleeve off the shaft. Don't lose the thrust washer directly under it.

Now for the weights. If retained with C-clips, pop them off and remove the weights. If plastic clips are used, *don't attempt to remove them*. They look like flat washers with a fluted ID, or fingers, on their inside diameter. Ford dealers don't stock them, so if you remove them they'll break and you won't be able to replace them. If this sounds like bad news to you, it shouldn't. Centrifugal-weight-pivot wear is not a problem. When they

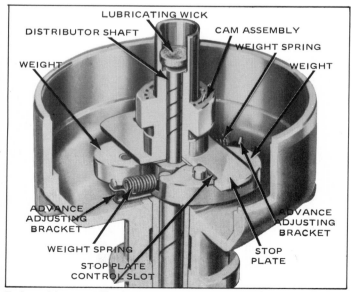 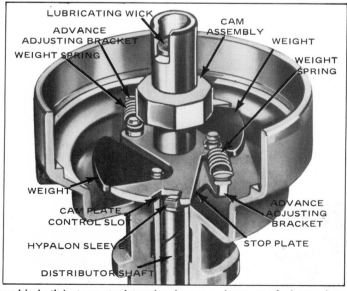

Early and late centrifugal-advance mechanisms. Weights, springs and slots are used in both instances to determine the rate and amount of advance. Late-style mechanism at right is used for both breaker-point and solid-state distributors. Drawings courtesy Ford.

Mark springs and weights so they can be reinstalled in their original positions before removing them. Don't overlook marking what they were mounted to.

Be careful when removing the centrifugal-advance return springs that you don't stretch them permanently. A pair of bent needle-nose pliers is handy here.

do wear the effect is minimal anyway. Therefore, just make sure the weights pivot freely.

Pre-1965 Cam, Spring and Weight Removal—The centrifugal weights in pre-1965 distributors are under the cam plate, and the springs are attached to the shaft plate and the cam. The cam is advanced by a pin on each weight which extends up through a slot in the cam plate. The cam must be removed before the springs or weights.

The same method is used for removing the cam as used in the late distributors. This will expose the springs and weights. Don't concern yourself with marking them, *just keep them together*. Lift each weight off its pivot and rotate the spring to unhook it from its mounting tab.

Breaker-point cams and armature sleeves are removed the same way. The retainer clip has to come out first. Doing this requires needle-nose pliers. With the clip out of the way, the cam assembly or sleeve should lift off easily, however it may need a little coaxing if the shaft is varnished up. Remove the thrust washer, too.

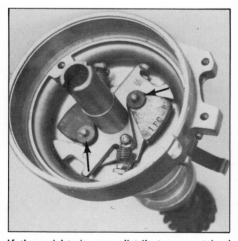
If the weights in your distributor are retained with this type of clip don't attempt removing them. You'll break them and they aren't serviced—this means you can't buy them.

Removing weights and springs together in a pre-1965 distributor. Lift weight off its pivot and rotate spring to unhook it. Weights and springs don't have to be marked if you do it this way.

Distributor-Shaft Removal—We're to the state of disassembly where all distributors are essentially the same. To remove your distributor shaft you'll have to remove the gear and collar first. Back up the gear hub—*not the teeth*—with the jaws of a vise or something that will clear the roll pin as it is being driven out. Use an 1/8-in. diameter pin punch or a drift punch that's no larger than 1/8 in., 3/4 in. from its end. Remove the pin from the collar also.

Press or Drive the Gear and Collar Off—If you don't have a press of some description to remove the distributor gear from its shaft, you'll have to use some less-sophisticated means to drive the gear off. Regardless of which method you use you'll have to back up the gear with something fairly substantial. Two parallel steel plates will do if you have a press, or some slightly opened vise jaws will suffice. Suspend the distributor upside-down by supporting it squarely on the backside of

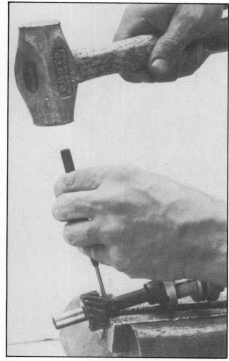
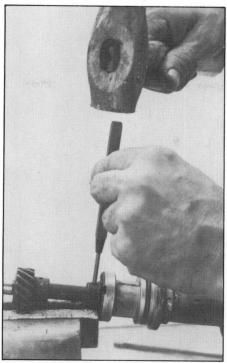

Support the gear and collar as I'm doing with this vise. I'm using an 1/8-in. pin punch to drive the roll pins out.

Support the housing on its base and over a hole or slot so the bushing will clear as it is being driven out.

To remove the top bushing the housing will have to be inverted and supported as shown in the sketch. It's advisable to remove this bushing with a press because of the additional force required, and the potential for housing damage. Regardless of how you remove it, press or hammer, you need a punch that's large enough to catch the bottom of the bushing, but no larger than 1/2-in. diameter, and long enough to reach down in the housing. Again, if you drive the bushing out, be careful!

Clean-Up and Inspection—Now that your distributor is disassembled, clean all the parts and inspect them closely. An old brush and a bucket, or can of cleaning solvent or lacquer thinner is perfect for doing this job. Be careful not to remove your locating marks on the cam, springs and centrifugal weights.

The shaft should look pretty good, but to give it a new look, polish it with some 400-grit emery cloth. Mike the worn and unworn diameters of the shaft and compare the results. If it's worn more than 0.001 in., replace it. As for the cam, feel the peak of the lobes across their width. If you feel any distinguishable ridge from the breaker-point rub block the cam

the gear teeth. Don't press or drive directly on the end of the shaft. Use a short section of your old oil-pump drive shaft inserted in the end of the shaft. As the shaft end nears the gear be ready to catch the distributor housing.

Remove the Bushing/s—If your distributor has the short lower bushing, you need a 3-1/2-in. long, 3/8-in. diameter punch to knock it out the bottom of the housing.

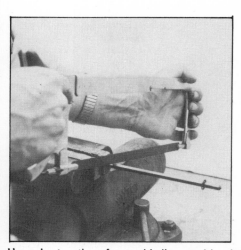

Use a short section of your old oil-pump drive shaft as a punch for removing the gear and collar from the distributor shaft. Make sure you support the backside of the gear like this and drive the shaft down through it. The collar isn't on as tight, so the housing can be used to back it up.

A long skinny punch being used to remove the bottom distributor-shaft bushing. Support the housing base and leave room for the bushing as it is driven out.

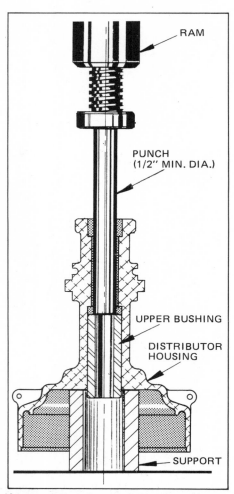

Using a press and a punch to remove the top bushing. A 3/4-in. deep socket supports the housing during this operation.

should be replaced. Otherwise, it's OK. Check the moveable breaker plate pivot and bushing if you haven't done so already. Loosely assemble the fixed and moveable plates and try to wiggle them. There shouldn't be any movement.

REASSEMBLING THE PIECES

After cleaning, inspecting and obtaining any parts you need for reassembling your distributor it's time to retrace your steps. Basically all that has to be done is to reverse the process, but with considerably more care.

Install the Shaft Bushing/s—Begin distributor assembly by installing the bushing/s in the housing. Install the bottom one first—Ford C5AZ-12132-A—by standing the housing upside-down and driving or pressing the bushing squarely in its bore using a 5/8-in. diameter punch. When the bushing feels solid as you're driving it in, it should be at the bottom of its counterbore and flush with the bottom of the housing.

Turn the housing over to install the top bushing—Ford B8HQ-12120-A. Lubricate the bushing and center it over its bore. Use a punch between the press ram or your hammer, and drive the bushing in *straight* until it stops at the bottom of its counterbore.

Burnish the Bushings—To make the bushing IDs as smooth as possible, they should be *burnished*. This is normally done with a *burnishing tool*, however these are not commonly obtainable, so use a 15/32-in. diameter ream. With the ream well oiled, turn it *backwards* in the bushings—opposite from the direction you turn it for reaming. This will smooth out any burrs or irregularities in the bushing bore/s.

Install the Shaft—Oil the distributor shaft and slide it into place. It should rotate freely. Secure it by sliding the collar—flanged end first—on the shaft and lining up the roll-pin hole with the one in the shaft. It's drilled off-center, so you may have to rotate the collar to get it to line up. If you are installing a new shaft you'll have to drill a 1/8-in. roll-pin hole. To set the correct end-play, drill a new hole *through the collar and the shaft* 90° to the original hole while a 0.024-in. feeler gauge is inserted between the collar and the housing, and the shaft is pulled down firmly in the housing. This will set the end-play. Drive the roll pin into place after you've lined up the holes.

Locate the bottom bushing squarely over its bore, then drive it in flush with the bottom of the housing.

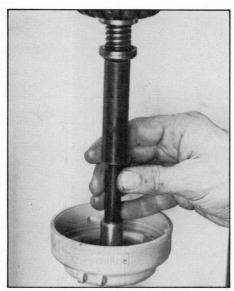  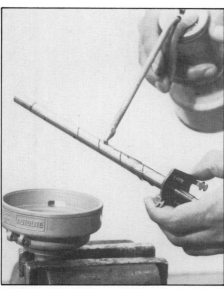

A press should be used for installing the top bushing. Oil the bushing OD and press it squarely into its bore until it bottoms in the counterbore. Use a punch under the ram of the press.

400-grit polishing cloth removes varnish buildup and restores the shaft's polish. Oil the shaft well before installing it.

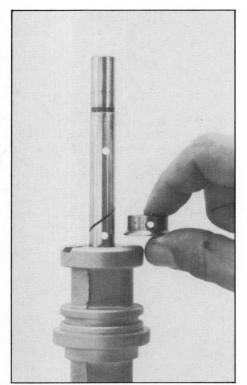 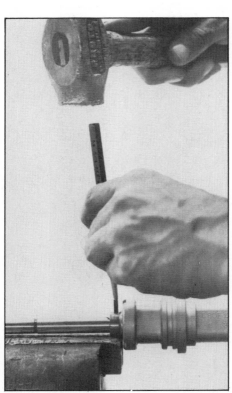 

Pin holes in the shaft, collar and gear are drilled off-center. Consequently, it is easier to line the holes up before sliding either the collar or the gear on the shaft. Once lined up, the pin can be installed while supporting the collar's backside. End-play should be no more than 0.035 in.

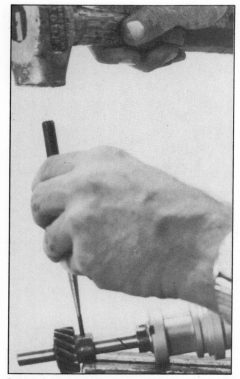
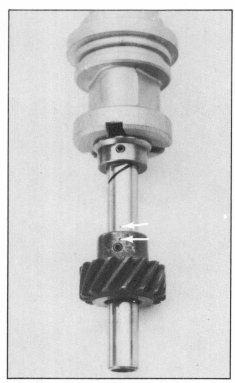

Support the backside of the gear when installing its roll pin. Prick marks (arrows) on the gear hub and shaft help line up the holes as you're pressing the gear on.

With the centrifugal weights installed on *their* pivots retain them with the C-clips. A finger on the clip is a precaution which helps keep from losing this elusive part.

Early-style centrifugal weights and springs being installed. Lubricate both pivots, hook free end of each spring over its tab and fit the weight into place. Lubricate the shaft and install the cam. Pins projecting out of the weights fit in the cam-plate slots.

Install the drive gear next. Start the hub end on first *after* lining up the roll-pin holes. It has to be lined up first because the gear will be too difficult to turn once it's on the shaft. Just as with the hub, if you've installed a new shaft you'll have to drill a new roll-pin hole. Do this after you've located the gear 4-1/32-in. below the bottom of the housing mounting flange—it's just above the O-ring groove—to the bottom of the gear. Install the roll pin and you're finished with the bottom end of your distributor.

Assemble the Centrifugal-Advance Mechanism—Gather up all your centrifugal-advance pieces in preparation for reassembling them. Have some moly grease on hand too. Start by spreading a *light film* of grease on the weight pivots, then install the weights using your marks to determine their locations. Secure the weights with C-clips.

Lubricate the upper end of the shaft by filling the grooves with moly grease, then slide the thrust washer into place over the shaft. Lubricate the bottom of the cam or sleeve and install it so the advance-limiting tab projects up through the correct slot at the edge of the cam or sleeve plate. Use your marks as reference. Check to make sure the Hypalon® sleeve is in place on the tab.

Again using your marks, install the advance-return springs between their shaft and cam or sleeve mountings. Attach them at the shaft plate first, then to the cam or shaft pin while being careful not to stretch the springs.

Complete the assembly of the cam or sleeve and centrifugal-advance mechanism by installing the retaining clip on the end of the shaft. Feed it in between the cam or sleeve and the shaft using your small needle-nose pliers, then push it into place using two small screwdrivers.

As for earlier distributors, assemble them using the same procedure, except install the cam and its retaining clip *after* you have its centrifugal weights and return springs in place and secured.

Breaker-Point Distributor Assembly—Base Plate Up—Assemble the moveable breaker plate to the base plate after lubricating the pivot with moly grease and making sure the three nylon wear buttons are in place. Install the spring washer so its curved section is against the top side of the bushing and the outside edge bears against the underside of the flat washer. Secure the plate, spring washer and flat washer with the C-clip.

Install the breaker-plate assembly, moveable or fixed, into the distributor housing. Rotate it until it lines up with the mounting holes in the housing, and secure it using the two mounting screws. The one closest to the primary-wire hole also attaches the breaker-plate ground cable.

Finish assembling the top end of the distributor by installing the primary wire, new breaker points and condensor. Set the points once they are installed by

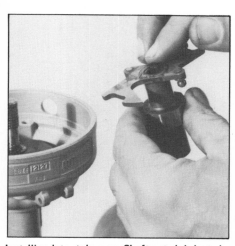 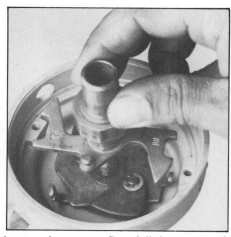 

Installing late-style cam. Shaft gets lubricated and thrust washer goes on first. A little grease on the cam base and it can be fitted between the weights. Make sure the right centrifugal-advance-limiting slot is around the tab. Hook the springs to *their* anchors, shaft end first followed by the cam end.

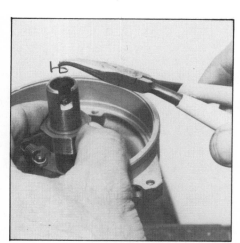 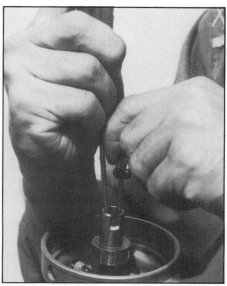 

Cam or sleeve-retaining clip is easier to install than it was to remove. Place it over the end of the shaft and push it into place using two small screwdrivers.

Don't forget the Hypalon® advance-limiting sleeve. Slide it over the tab so it locks in place below the tab ears.

 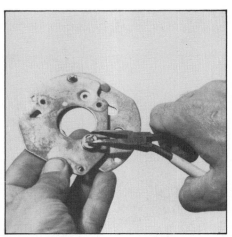 

Assembling the moveable breaker plate to its base plate, first lubricate the pivot and make sure the three nylon buttons are in place. Secure the breaker plate with the spring washer, flat washer and C-clip. Spring washer convex side should be against the breaker-plate bushing as shown at right.

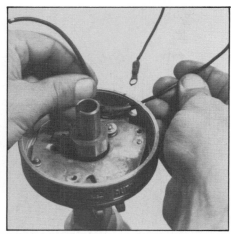

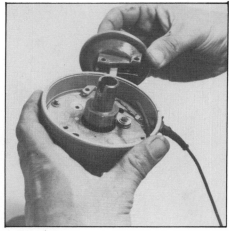

Secure breaker-plate assembly in housing using two screws. Ground strap is installed under the screw closest to breaker-plate pivot as pictured. Feed the primary wire through the housing and fit its molded section into place. Put vacuum-advance can and link in place, securing it with two screws at the housing and a C-clip at the plate.

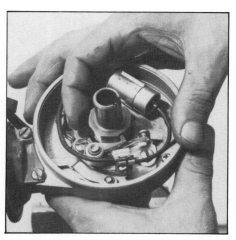

Finish assembling breaker plate by installing points and condenser. Lightly lubricate cam and adjust the points with a clean feeler gauge.

Don't forget the cam or sleeve lubricating wick. Push it down on top of the shaft and saturate it with light oil.

rotating the cam so the point's rub block is on the top of one of the lobes. Adjust the earlier style distributor's point gap to 0.015 in. and the later ones to 0.018 in. The HP289 and Boss 302 dual-point distributors are set to 0.020 in. When setting the points, make sure your feeler gauge is free of oil or grease. Recheck point gap once the points are set and secured to make sure it didn't change when you tightened the attaching screws. Finish the job by *lightly* lubricating the cam with the grease that accompanied your new points, then install the rotor and cap.

Breakerless Distributor Assembly—Base Plate Up—Install the base plate using one screw opposite the vertical slot in the housing. Install the magnetic-pickup

Install rotor and cap to keep innards clean while distributor is waiting to be installed in its engine.

After installing solid-state distributor's base plate, lubricate the pickup pivot. Position pickup assembly on base plate and fit harness molded section into housing slot, securing it with a base-plate screw. Secure pickup to base plate with wire clip.

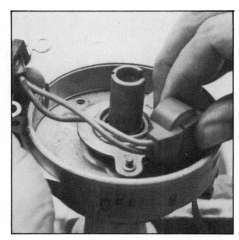

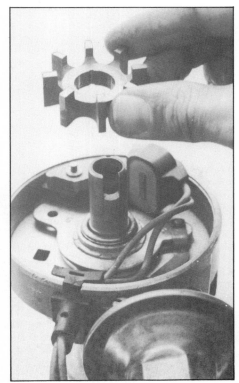

With the grooves in armature and shaft lined up and armature vanes pointing up, push armature down on shaft until it bottoms. Lock it to the shaft by installing roll pin in groove. I'm tapping it into place with a hammer and punch.

assembly after lubricating its pivot, and retain it with the wire retaining clip. Slide the wiring harness into its slot in the housing, then secure it and the base plate to the housing with the remaining screw. Install the vacuum-advance diaphragm by by securing its link to the magnetic pickup with the C-clip and bolting the vacuum can to the distributor housing with the two long screws.

With the vanes pointing up, install the armature on the sleeve. Push it all the way down on the sleeve, lining up a longitudinal groove in the armature and the shaft. Then install the small roll pin in the groove. Tap it into place using your hammer and punch. Install the rotor and distributor cap and the distributor is ready to be installed in the engine.

TESTING AND CALIBRATION

After your distributor and engine are installed in your car or truck, you'll want to make sure they will perform at their peak—maximum mileage and power with a minimum amount of pollutants. A distributor is one of the major factors in determining an engine's performance, and to tune one to its full potential requires expensive and sophisticated tuning equipment in the hands of a tuning specialist. This is the subject of the last chapter.

9 Engine Installation

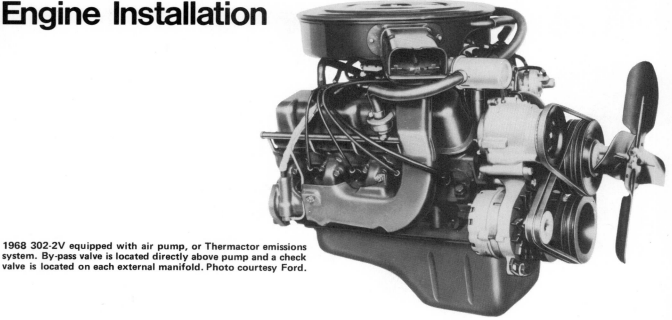

1968 302-2V equipped with air pump, or Thermactor emissions system. By-pass valve is located directly above pump and a check valve is located on each external manifold. Photo courtesy Ford.

You're now ready for the last big operation, putting your newly rebuilt engine back where it came from—into the engine compartment. Consequently, you have to get those boxes of bolts, nuts and parts out from where you've had them safely stored. You'll also have to round up the engine hoist, jack and jack stands.

Before you begin the installation process, consider cleaning the engine compartment, accessories, related brackets and hardware so the total job under the hood will look good. Cleaning the engine compartment requires a spray-can of engine cleaner, a stiff bristled brush for the stubborn grease and dirt and a garden hose. Don't forget to protect your car's body paint. Engine cleaner is strong stuff. You can use the same process to clean all the bolt-on parts and accessories, or you can load them up for a trip to the local car wash. The advantage is you leave the mess there and the hot high-pressure water makes the job a whole lot easier. Let's get on with installing your engine.

TRANSMISSION FRONT SEAL
Any leakage into the clutch or converter housing which obviously didn't come from the engine should be corrected **now**. This is usually accomplished by replacing the transmission's front seal, regardless of whether it is an automatic or standard transmission.
Automatic Transmission—If you have an automatic transmission, replace the front-pump seal whether it shows signs of leaking or not. Chances are, it will start leaking before you get out of the driveway if you don't replace it now. To replace the front-pump seal or an automatic transmission, you'll have to remove the converter first. The converter can be drained before or after it's removed. This will prevent spilling fluid while reinstalling it. It's best to drain it now and make up the difference later by filling the transmission after you get everything back together. To drain it before removal, position one of the drain plugs at the bottom by rotating the converter. Position a bucket directly underneath *before* removing the plug. After the fluid has drained, replace the plug and remove the converter simply by pulling on it, but be ready to handle some weight, it is not light. Now, if you decide not to

Replacing automatic-transmission front seal requires removing the torque converter. Slide it off the transmission input shaft and drain it.

drain the fluid first, pull the converter off and tilt it front down so the fluid doesn't run out. Now that you have the converter off, drain it by tilting it the other way over a suitable container.

Removing the converter exposes the front-pump seal. To remove it, put a screwdriver behind the seal lip and pry it out. Go around the seal and pry a little bit at a time and it will eventually pop out. Some people remove the seal with a chisel and hammer, driving against that little bit which extends beyond the front-pump bore. I don't recommend this because of the risk of damaging the *stator support* or the seal bore. So pry it out.

Before installing the new seal, clean the front-pump bore by wiping it clean of any oil using a paper towel and some lacquer thinner. Now, run a small bead of oil-resistant sealer around the periphery of the new seal, preferably on the edge that engages the pump bore first so the sealer will wipe the full face of the mating surfaces as it is installed. This makes a more effective seal. Just before installing the seal, wipe some clean oil on the lip of the seal so it will be prelubed. To install the seal, place it squarely in the front-pump bore and tap lightly around it with a hammer. Be careful it doesn't cock in the bore. Keep doing this until you feel the seal bottom firmly all the way around. Wipe any excess sealer from the outer edge of the seal and you're ready for the converter. While you're still in the engine compartment, clean the inside of the converter housing so it will start out like new. Make sure the front face of the housing—the surface that mates with the engine—is clean and free of any burrs or nicks. This will ensure a good fit between the engine and the transmission.

Before installing the converter, check

134

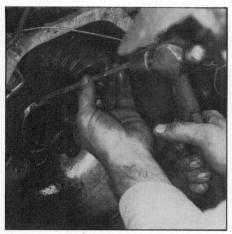  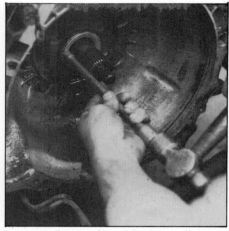

Crude, but effective way of removing the front transmission seal—an old screwdriver and a hammer. Apply a bead of sealer to the OD of the new seal and install it squarely in its bore. A long extension and hammer are being used to tap it into place.

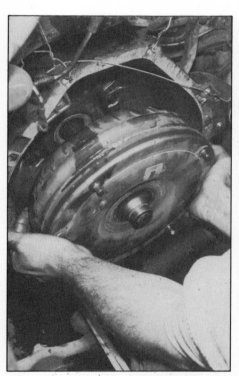  

After oiling the converter's oil-seal surface, lift it into place. Spin the converter while pushing on its center to engage it with the input shaft and the front pump. With the converter fully engaged, locate the marked stud directly at the bottom. Wipe the housing mounting face clean and the transmission is ready for the engine.

the outer surface of the spline, or the surface which turns against the seal that you just replaced, particularly if the one you replaced leaked. If it appears to have any nicks or burrs or is at all rough, polish it with some 400-grit paper. If you don't the seal will quickly be destroyed. If the surface appears OK, just clean it with a rag and some lacquer thinner to remove any deposits such as varnish.

Now, you're ready to install the converter. Lightly oil the outside of the surface you just cleaned. It's good insurance for the life of the pump seal even though you should've already oiled the seal. It will also reduce the friction on the seal when you install the converter. Start the converter on the transmission-input shaft and rotate it back-and-forth while pushing on it until it engages with the transmission. Be careful here because the converter must engage not only the transmission-input shaft, but must also key into the front pump and the converter-support. As a result you will definitely feel two engagements, and maybe all three. When you are certain the converter is on all the way, position the marked converter-to-flex-plate mounting stud directly at the bottom of the housing in preparation for installing the engine.

Standard Transmissions—It's rare when a manual transmission leaks into its bellhousing, but when it does the culprit is usually the gasket between the transmission case and the bearing retainer. If you find your transmission is not leaking, leave it alone because chances are it will never leak. On the other hand, if there

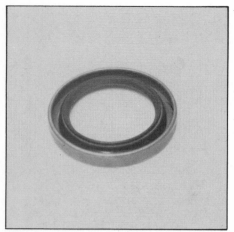

Manual transmission input-shaft seal. Install seal with lip pointing toward transmission. Use a socket the diameter of the seal for driving it into place.

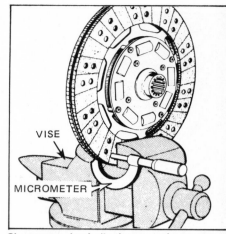

Clamp your clutch disc *just enough* to compress it for measuring its thickness and no more. If your disc measures near 0.280 in. consider it worn out and replace it. Photo courtesy Schiefer.

are signs of transmission grease in the bellhousing or on the clutch, see whether the grease is coming from the gasket or from around the input shaft. Grease may be running out the front of the retainer—the end that the input shaft comes out. If your car is equipped with the *top loader* 4-speed transmission, the input-shaft seal will be leaking. To replace it, remove the retainer, support it on a solid surface and drive the seal out with a long punch and a hammer by gradually working around the back of the seal. It will eventually come out, but don't be impatient by trying to remove it in one whack. When installing the new seal, turn the retainer over and set the seal squarely in its bore. Use a socket or a piece of steel, brass or aluminum which approximately matches the outside diameter of the seal—a little larger won't hurt, but smaller may damage the seal—and a hammer to drive the seal in. After the seal is in place, lubricate its lip with some clean oil and replace the bearing retainer on the transmission with a *new* gasket and torque the retaining bolts: 19—25 ft. lbs. for side-cover 3-speeds, overdrive 3-speeds and top-cover 4-speeds; 12—15 ft. lbs. for side-cover 4-speeds; and 30—36 ft. lbs. for top-cover 3-speeds.

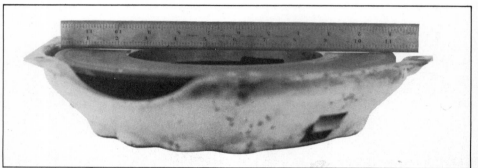

Two pressure plates suffering from having been over-heated. One is warped as indicated by the gap between ruler and pressure-ring surface. Heat checks and chatter marks on other pressure-plate's surface indicates it has been extremely hot. If your pressure plate is suffering from either of these symptoms, replace it. The same applies to your flywheel.

ENGINE PREPARATION AND INSTALLATION

Now turn your attention to the engine and the components which must bolt to it prior to installation.

Clutch & Flywheel—If you have a standard transmission, now's the time to find out whether or not your clutch or flywheel needs reconditioning or replacement. This is, of course, purely academic if the clutch slipped before. You'll already know something has to be replaced. The disc for sure, and maybe the pressure plate, or even the flywheel. On the other hand, if it didn't slip, the way to gauge clutch wear is by the thickness of your clutch disc. As a disc wears it naturally gets thinner. If not corrected in time this eventually results in pressure-plate and flywheel grooving as the rivets will contact them directly.

Measure the Clutch Disc—To measure a clutch disc, you'll have to compress it to its engaged thickness. When doing this be careful not to damage the facings or get grease or oil on them. Compress the disc with some sort of clamp. A vise, C-clamp or Vise-Grip® pliers will do. Just clamp the disc enough so the wave spring (marcel) is flat. Measure the disc thickness as close as possible to the clamp to get an accurate reading. New disc thickness is about 0.330 in. which allows for 0.050 in. of wear, leaving a worn-out disc at 0.280-in. thick. At 0.280 in., the rivet heads are

Before installing your flywheel or flexplate, install the engine plate. Use gasket adhesive to hold it in place. Don't forget the load-spreader ring if you have a flexplate. Apply sealer to the bolt threads, then install and torque the bolts 75–85 ft.lbs.

An old transmission-input shaft being used to align the clutch disc with the crankshaft pilot bearing. It can be removed after the pressure plate is secured by tightening each bolt a little at a time.

even with the friction surface of the facing and ready to damage the flywheel and pressure plate. As for what thickness you should replace it, I don't know except that if it only shows 0.010-in. wear, it obviously doesn't need replacing. However, if it's down to 0.290 in. thick, replace it. Anything in between depends how long you want to drive before you have to replace your clutch.

Pressure-Plate Inspection—Assuming you have the disc problem all sorted out, inspect your pressure plate. The general rule here is, if the disc wore out in a short time from one thing or another, the pressure plate may require replacing, but if the disc looks good or gave you many miles of good service, chances are it's all right. The first thing to do in determining whether a pressure plate is OK or not is by just looking at it. If the friction surface is bright and shiny, and free of deep groves which result from the rivet heads of a more than wornout disc, the pressure plate is probably OK. If the friction surface has black and blue spots (hot spots) covering it, it has been slipped excessively and may be full of heat checks (cracks). In any case, check the pressure-plate's friction surface by laying a straight edge across it. If it is warped, it will show an increasing air gap between the straight edge starting from nothing at the outer periphery and increasing to the inner periphery of the pressure plate—it will be warped *concave*. If the hot spots, heat checks or warping are not too bad, you should compare the price of resurfacing your pressure plate versus purchasing a new or rebuilt one if your budget is tight. Otherwise don't waste time—replace it. Regardless, if your clutch has seen more than two discs it should be replaced no matter how good the friction surface looks. The bushings in the release mechanism will be worn and the springs which apply the pressure plate's *pressure* will be weak from fatigue, resulting in reduced clutch capacity. You may have gotten away with it with a tired engine, but now the engine will be developing more torque than when it was new.

Flywheel Inspection—Because the flywheel friction surface sees the same slipping as the pressure plate, it will suffer from the same basic ills. If your pressure plate has hot spots, heat checks or is warped, then look at your flywheel. Because it has more mass than the pressure plate it will absorb more heat, meaning it won't receive the same amount of damage. For instance, you won't see a warped flywheel because the heat required to do it would destroy a clutch's facings first, but on the other hand it is quite common for a flywheel to get hot enough to show hot spots and heat checks. Also, a flywheel is not immune to rivet damage, it it can be grooved as easily as a pressure plate. If your flywheel has hot spots, look further for heat checks. If your pressure plate is grooved from a worn disc, chances are the flywheel is too. If it is heat-checked or grooved very badly, it should be resurfaced. Your engine machine shop can do this or recommend a place to get it done. If you do have a flywheel resurfaced, make sure the job is done on a grinder and not on a lathe. Hot spots on the friction surface of a flywheel are very hard causing the surface to have *bumps* on it if surfaced on a lathe, whereas grinding will make it flat. Finally, if your flywheel has heat checks that have developed into long radial cracks (like the spokes of a wheel) toward the center, *replace it*. It can't be satisfactorily repaired by resurfacing. More important, it's dangerous! A cracked flywheel is highly susceptible to coming apart and *exploding* like a hand grenade.

A flywheel that appears to be all right should still be cleaned. For instance, the clutch friction surface will be shiny, or may have some deposits on it. To help it and the clutch, use fine-grit sandpaper—400 grit will do—to roughen the surface and to remove any resin binder deposited by the clutch facings. Follow this up with a cleaning using a non-petroleum-based solvent such as alcohol or lacquer thinner. This will remove any oil deposits. Give the same treatment to the pressure plate if you are planning to use your old one again.

The Clutch Release Bearing—The final customer to look at is the release bearing. It should still be on the end of the release lever in the clutch housing if you didn't remove it. The rule is: replace it if you are replacing your clutch. The next rule is replace it anyway. The relative cost of a new bearing and the amount of work it takes to replace it now as opposed to when your engine is back in should give you a good idea as to what you should do. Regardless of whether you replace the bearing or not, grease the inner bore of the bearing hub. There's a groove in the hub for just that purpose. Don't overdo it because too much grease can result in clutch failure if it gets on the clutch disc. Just fill the groove with some moly grease and replace the bearing assembly on the lever.

Engine Plate—Whether you have an automatic or a standard transmission, there's an engine plate to install before the flywheel or flexplate can go on. Because it depends on the bellhousing or converter housing to hold it to the engine, it'll be constantly falling off while you are trying to install the flywheel or flexplate and will fall away from the rear face of the engine block when you're attempting to button the engine up to the transmission. Consequently, a little tip is in order here. Apply a thin bead of 3M® weatherstrip adhesive to the rear face of your engine block. It *will* hold the engine plate in place. When installing the engine plate, fit it over the locating dowels with the starter hole to the right and hold it in place for a minute or so to make sure the adhesive gets a good grip on the plate.

Flywheel or Flexplate—Make sure the crankshaft-mounting flange is clean and free of burrs or knicks which can cock the flywheel or flexplate after it's mounted, causing it to wobble. This is particularly important in the case of flywheels because it can cause clutch or transmission problems. Install the flywheel or flexplate using the six grade-8 bolts with a drop of Loctite Loc N' Seal® on the threads. *Don't forget the sealer.* The mounting holes in the crankshaft flange are drilled and tapped straight through the flange into the engine's crankcase, and as a result crankcase oil will leak past the threads into the bellhousing or converter housing if they aren't sealed. For automatic transmissions, this means lost oil and spots on the driveway. For standard transmission, it also means an oiled clutch—consequently a ruined clutch. When locating the flywheel or flexplate on the crankshaft flange, rotate it until the holes in the crank and the flywheel or flexplate line up. The reason for this is the flywheel/flexplate must be mounted in a certain position on the crank to maintain engine balance. If you are mounting the flexplate, you will also have what amounts to a *load spreader* which is a ring with the same bolt pattern as the flywheel. Its function is to spread the load exerted by the mounting bolts so the area around each bolt head doesn't become overstressed. This avoids cracking or breaking the flexplate at the mounting bolts. Don't forget this little ring with the six holes in it. Snug the bolts up and torque them in a zig-zag pattern to 75–85 ft. lbs.

Mounting the Clutch—Now that you have the flywheel mounted on the engine, you're ready for the clutch. Remember to avoid touching the friction surfaces of the pressure plate, the disc or the flywheel. Grease on the clutch can cause bad things like grabbing or slipping to happen, so be careful what you touch. To ensure the clutch friction surfaces are free from oil and grease, clean with lacquer thinner even though you're *sure* you didn't get any oil or grease on them. To mount the clutch you'll need one tool to line the center of the clutch disc up with the crankshaft-pilot bearing. An old transmission input shaft works well, or there are special tools for this job which are inexpensive. Although you can probably do this job yourself, a friendly third hand comes in handy. To install the clutch, hold the pressure plate and the disc against the flywheel while starting at least 2 pressure-plate mounting bolts and their lock washers. As soon as you get the 2 bolts started to hold the pressure plate and the disc in place, install your clutch-alignment tool in the center of the disc and into the crankshaft-pilot bearing. You can now install the rest of the bolts and washers *loosely*. A word of caution at this point—don't tighten any of the pressure-plate mounting bolts all at once. This will result in junking the pressure plate because it will bend the cover. Tighten each bolt a couple of turns at a time going around the pressure plate till the cover is firmly against the flywheel. Now you can remove the alignment tool and torque the pressure-plate bolts 12–20 ft. lbs.

Install the Carburetor Heat Riser—If your engine is equipped with a heat-riser shroud and/or a spark plug shroud, install them on the exhaust manifold/s. Fit them over the exhaust-manifold-bolt studs and secure them in place with the mating nuts.

Spark Plugs—Now seems like a strange time to install a new set of spark plugs. You may think there is a good chance some may get broken during the installation process, but fear not. The exhaust manifolds do an excellent job of protecting the plugs. Remove the old plugs which you installed to keep the cylinders clean and dry, and replace them with the new ones. Check them for proper gap and set them accordingly. Torque all the plugs 15–20 ft. lbs. except for the one in the number-one cylinder. Just snug it up. It has to be removed when you install the distributor.

Engine Mounts—The engine mounts used with the small-block Ford vary according to the car they are used in, however their basic design remains the same. All use the single through-bolt for ease of engine installation and removal. Some mounts are not interchangeable from side to side, so make sure you identify which is which. When assembling the mounts to the engine, make sure you have all the parts that go with each mount. For instance, the '68 Mustang pictured here uses a starter-cable bracket which attaches to the right-front engine-mount bolt. After snugging the bolts up, torque them to 40 ft. lbs.

Alternator Bracket—If your engine only has an alternator mounted on its right side of it it has an air pump in addition the the alternator, but with the alternator mounted above the air pump, now is a good time to install the alternator bracket. It makes a good handle for guiding the engine into place. Install the bracket loosely to the water pump and then install the long alternator pivot bolt which goes through the bracket, the alternator and then threads into the head. Run the bolt into the head a few turns until the bolt stabilizes, then tighten the bracket mounting bolts at the water pump to 12–15 ft. lbs. The reason for installing the alternator-pivot bolt is to align the bracket so the alternator and bolt will assemble later on without binding. Leave the bolt in the head until you're ready to install the alternator.

Oil-Pressure Sending Unit—You'll have to use your best judgement as to whether or not to install your oil-pressure sending unit now, or after your engine is in place. Base your decision on whether you have the smaller warning-light type or the large sender and extension as used with a gage. Remember, the extension is easily broken if it is bumped during the engine installation.

To install the warning-light sender, thread it into the lower-left-front side of the engine block after coating its threads with sealer. You'll need a 1-1/16-in. open-end wrench to turn the large hex. There is a special socket made just for this sending unit, however they aren't normally found in even the most professional of tool chests, so count on using an open end. Tighten the sender so it is snug.

You'll have to use your discretion with the larger gage-type sender. Whether you decide to install it now or after your engine is in place, use sealer on the extension threads and tighten the assembly with an open-end wrench. Use the wrench at the cylinder-block end of the extension to minimize the bending load on the extension. Turn the extension and sending unit so the assembly is snug and *the sender is pointing up*.

Fuel-Pump—To install your fuel pump, you'll need its gasket and 2 attaching bolts 3/8-16 x 1-1/2-in. long. Use sealer on both sides of the gasket and coat the end of the pump actuating arm at the cam bearing surface with moly grease. Install the pump by forcing it to line up with its mounting holes as the actuating arm contacts its cam, then thread the bolts into place. Torque them 20–25 ft. lbs.

Crankshaft Pulley—Install the crankshaft pulley on the crankshaft damper. For good centering of the pulley it has an *extruded* hole which pilots into the center of the damper. Make sure the pulley is piloting before you tighten the bolts. Torque them to 40–50 ft. lbs.

Position the Crankshaft—There's not much more you can do with the engine out, so the big moment has come. Attach a chain to the front of the left head and to the rear of the right head as you did when removing the engine. Now, if you have an automatic transmission, the most important thing to do is to rotate the crankshaft so the marked hole on the flexplate will be at the bottom so it will line up with the torque converter. It won't take long for you to find out how important this is if the converter studs and the flexplate holes don't line up.

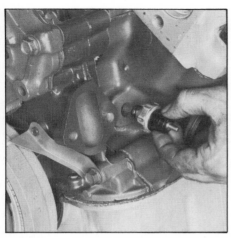  

Use sealer on the oi-pressure-sending-unit threads. Short "idiot-light" sender can be installed now without danger of damaging it during engine installation, but not the gauge type. It is shown being installed now only so you can see how to do it once the engine is installed. When installing the gauge-type sender, rotate it so the end points up as shown.

With this in mind, double-check to see if the converter is positioned correctly and you should be ready to get the engine and car back together again.

INSTALLING THE ENGINE

Lift the engine and position it so it's level side-to-side and level front-to-back or a little low at the rear. Set the engine down and adjust it till you get it right. It'll save a lot of unnecessary trouble when you're trying to mate the engine to the converter or bellhousing. Now is another one of those times when it's handy to have a helpful friend on the scene—someone to help guide the engine into place and get those first couple of converter-housing or bellhousing bolts started. Position the floor jack under the transmission and raise it up enough so you can remove the wires or whatever you used to support the transmission and exhaust pipes. If you have a standard transmission, put a *few* drops of oil in the crankshaft pilot bearing and smear a *little* moly grease inside the clutch disc splines and put the transmission in gear. Take one last look in the engine compartment and make sure everything is tied up out of the way so nothing will interfere with the engine as it is being lowered into place.

Lower the Engine in—Position the engine over the car or the car under the engine and lower the engine gently into the engine compartment. The front of the transmission will need to be raised higher than it would normally be in its installation position so the engine can be lined up and engaged with it while clearing the engine mounts. This is particularly true with a manual transmission. For a manual transmission, carefully guide the engine back, centering the clutch disc on the transmission-input shaft. When the

Install the fuel pump. Use sealer on the gaskets and torque the bolts 20—25 ft.lbs.

disc and input-shaft splines contact each other, rock the engine back-and-forth while holding the engine against the input shaft until the splines engage. You can then put the transmission in neutral, leaving the engine free to rotate into engagement with the bellhousing. This means lining the engine dowel pins up with the bellhousing holes. When this happens, the two will literally "click" together, signaling you to get at least one bolt in on each side of the housing in as low a position as you can reach from above.

When it comes to automatic transmissions, the converter studs and the engine dowels must engage almost simultaneously. Do just as is done with a standard transmission and push on the engine while rocking it, but not as much. If you have things lined up well, this job will only be difficult. You'll have to climb right in there with the engine, even to the point of standing on your head so you can look down between the rear-face of the engine and the converter housing. This is so you can see the relative positions of the two. When the engine and the housing come together, get a bolt into the engine on each side and as low as you can reach. Tighten the bolts.

Engine-Mount Bolts—Now that the engine and transmission are mated, you can remove the jack from under the transmission. Lower the engine while lining up the holes in the engine mounts. Because the right one is more accessible, install one of the through bolts and the other mount should line up automatically. Before putting the nuts on the bolts and tightening them up, lower the engine all the way down and remove the lifting chain from the engine. You're now finished with the engine hoist. Jack the car up in the air and set it on jack-stands. With the car firmly supported, slide underneath and finish installing the engine-mount through bolts. A box-end wrench on the head of the bolt to keep it from turning and a *long* extension with a socket and ratchet on the nut is the easier way to tighten these.

With the flexplate (automatic transmission) mark positioned at the bottom and all the hardware you can hang on your engine, it is ready to be reinstalled. Hoist it over its compartment and do as Ron is saying, "Let 'er down easy, Sam." You'll need a jack under the transmission so you can line it up with the engine. It takes a lot of joggling to get a couple of bolts in the top of the housing.

Bellhousing or Converter Housing to Engine—Install as many bellhousing or converter-housing-to-engine bolts as you can manage from under the car. Torque them 40—50 ft. lbs.

Starter Motor—Make sure that when you're gathering the starter motor and the attaching bolts that you get the right two bolts. Many, but not all, starter-motor mounting bolts are special and can be recognized by their reduced hex size and the integral washer forged into the bolt. The normal hex size for a 7/16-in. bolt is 5/8-in. whereas these are 1/2-in. The reason is to provide wrench clearance between the bolt and the starter motor—the bolts are *that* close!

When putting the starter in place for mounting, cars such as the Falcon, '72 and earlier Mustang/Cougar, all Maverick/Comet and the Granada/Monarch require that the end opposite from the geared end be guided up behind the steering linkage, slipped forward and then backed into the engine-plate opening. It's not difficult to do, things just have to be done a particular way. Once you get the starter in place, thread the top bolt into the bellhousing/converter housing first. This will support the starter so you can

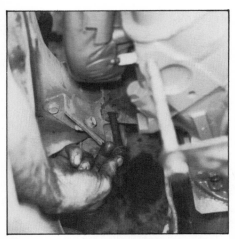
With the engine and transmission engaged lower the engine until the holes in a mount line up. Install the through bolt then line up the other mount. Bolts can be secured with their nuts after you have your car up in the air.

Connect the two leads to the block with the *star washer against the block* to ensure a good ground. With a friend at the ignition switch or with a remote starter switch, you can now turn the crank the modern way. If you have a standard transmission, you don't have to worry about it.

Converter-to-Flexplate Nuts—For an automatic transmission, the converter can now be attached firmly to the flexplate. Remember you'll have to tighten one, then rotate the crank approximately 90° to get to the next one until you have nuts on all studs. Be careful again if you are using the friend on the ignition switch approach. Make sure he understands the switch shouldn't be touched *unless you say so*. Torque the nuts to 25 ft. lbs. Install the converter cover to the housing. If the upper lip of the cover fits a little tight between the engine plate and the converter housing, tap it lightly into place with a hammer. When lined up, install the two lower bolts in addition to the one at the top opposite the starter which goes through the engine plate, the converter cover and then threads into the housing. Maximum torque is 15 ft. lbs.

Engine-Plate-to-Bellhousing Attachment—The engine plate used with standard transmissions extends down in front of the bellhousing to seal the clutch area. Two bolts clamp the engine plate to the bellhousing. Torque them to 15 ft. lbs.

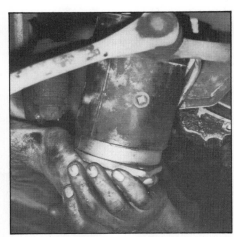
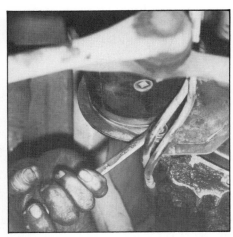
Make sure you have the small 1/2-in. hex-head bolts before you install your starter. After you jockey the starter into place, install the bolts and torque them if you find room for your torque wrench—20 ft.lbs. max. A "calibrated hand" and a box-end wrench are being used here.

release it. Be careful when tightening these special bolts. It's easy to round their heads off. Torque them to a maximum of 20 ft. lbs.

After getting the starter motor into place, connect the battery hot-lead. Route the cable through any clips or brackets it may run through as it goes to the battery such as the bracket used at most right-side engine mounts.

Battery and Alternator—If you want to use the starter to rotate the crank when you're fastening down the torque converter to the flexplate on an automatic transmission, you'll have to install the battery, connect it to the starter-solenoid lead, connect the engine ground cable to the engine and the battery. Because the battery ground and the alternator/generator lead connect to the engine, hang the alternator/generator on the engine loosely. Don't forget the spacer if yours uses one.

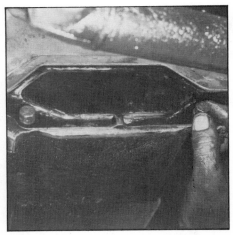
Install the battery, alternator/generator and their leads now so you can bump your engine over to expose the converter studs. Install the nuts and torque them to 25 ft.lbs. Install the cover plate using the two bolts with lock washers.

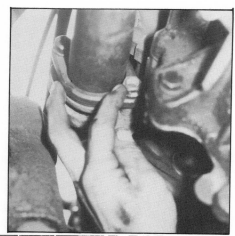
Using new gaskets, hook up the exhaust-inlet pipes to the exhaust manifolds. As you look up at the left pipe, you can see this area is tight.

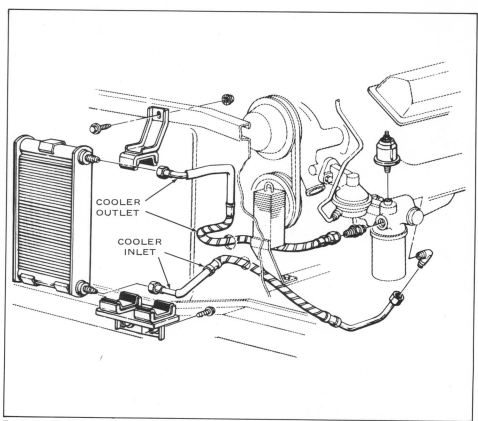
For Boss 302 engines connect the engine-oil cooling lines to the oil-filter adapter and install the oil-pressure sending unit. Regardless of which engine you have, install its oil filter now. Drawing courtesy Ford.

Before reinstalling your carburetor you may want to rebuild it using a standard rebuild kit. Have some lacquer thinner handy for cleaning the parts and follow directions in the kit.

Exhaust Pipes to Exhaust Manifold—If your gasket kit contains two conical-shaped pieces of asbestos with approximately a 2-in. hole in each, they are exhaust-pipe-to-manifold gaskets. Remove the old ones from the exhaust-pipe ends and slide the new ones on. You can now pilot the exhaust pipes into the exhaust manifolds, slide the flanges up each pipe and hold each in place while you get a nut on one of the manifold studs. Long and agile fingers are a definite asset for this operation. One nut on a stud will free you so you won't have to hold the pipes in place. You can now install the other nuts and torque them to a maximum of 35 ft. lbs. Some of the earlier engines, the 221s and the 260s, used direct-mating flanges which require flat gaskets. If yours is this type, apply some weatherstrip adhesive on one side of the gaskets to hold them in place, install the gaskets, lift the pipes up to the manifold and install the nuts. Torque them to 35 ft. lbs.

Oil Filter—Before setting the car down for the last time, install the filter. It's usually easier from under the car. Prior to installing it, fill the filter about half full of oil so the engine will receive lubrication sooner—it won't have to fill the filter first. Smear oil on the seal and thread the filter on. Turn the filter 1/2-turn after the seal makes contact with the engine block. If you have a Boss 302 connect the oil-cooler lines to the back of the adapter and the top one to the front. Check to make sure the oil-pan drain plug is tight and you can set your car down.

Engine Oil—If for no other reason than to take a break, now's a good time to fill your crankcase with oil. Because your oil filter is partially filled, add one less quart than specified—5 quarts for all passenger cars and 6 quarts for *most* light trucks. Check the oil level after running your

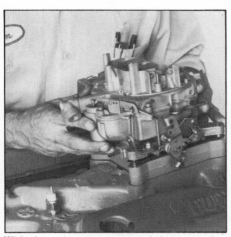  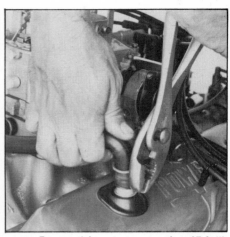

With the carburetor spacer and the two gaskets in place, install your carburetor. Use washers under the nuts. Do not tighten nuts more than 15 ft. lbs. Make the necessary connections to the carburetor now: fuel line, PCV hose and choke heat tube. Install a new fuel filter.

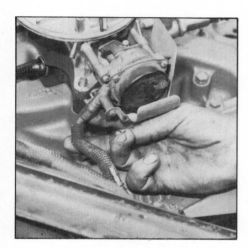

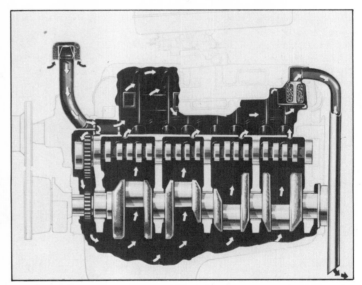

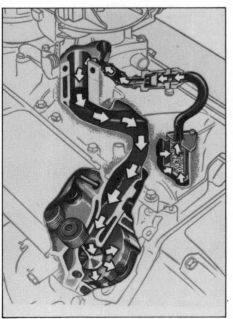

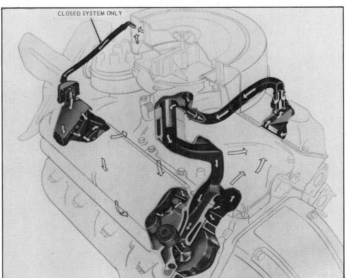

Crankcase-venting systems from the early road-draft tube and the early-style PCV system to the late closed PCV system. No matter which system your engine uses, clean the oil-filler cap by soaking it in solvent to let the crankcase breathe. Drawings courtesy Ford.

engine, then bring it up to the full mark.

Carburetor—Carburetors have varied widely over the years during the existence of the small-block Fords—from the performance-oriented high-performance 289 with its four-barrel carburetor and manual choke to some of the later emissions and mileage-oriented two-barrel equipped 302s and 351s. Regardless of the engine, most use a spacer between the carburetor and the intake manifold. This insulates the carburetor from the intake manifold and provides an inlet for the PCV system into the engine's induction system. Install a gasket on the manifold followed by the spacer and then another gasket. You can now position the carburetor on the manifold, or spacer and secure it with the four attaching nuts. Go around the carburetor tightening each one *a little at a time* until all are at the correct torque. *Don't* exceed 15 ft. lbs., they are easily stripped. Progressively tightening the bolts eliminates the possibility of breaking the carburetor flange.

Crankcase Ventilation—Hook up the lines to the carburetor starting with the PCV valve and hose. First check the valve. Shake it and listen for a rattle. If it rattles, the valve is OK to install. Otherwise, soak it in solvent until it loosens up. Replace it if it doesn't rattle after soaking. Plug the valve into the valve cover, usually at the right rear. Connect the hose from the valve to the carburetor spacer. If you have one of the earlier small blocks, the crankcase portion of the PCV system ran to the back of the intake manifold and used a valve which spliced into hose. These can be found on some of the 221s, 260s and pre-1966 289s. Those engines not using a PCV system used a *road-draft tube* instead. It replaced the PCV system at the manifold and vented the crankcase to the atmosphere. If yours is this type, I suggest changing it over to the PCV system because of the emissions aspect and to keep the engine cleaner inside. It will also get slightly better gas mileage! The harmful fumes circulating in the crankcase will be drawn into the combustion chamber and burned rather than contaminating the crankcase oil or venting into the air.

Choke—If you have a high-performance 289, hook up the choke cable and adjust it. With the choke knob all the way in, clamp the cable conduit after securing the cable end to the choke-plate lever with the choke *open*. On the other hand, if you have an automatic choke, connect the heat tube from the right exhaust manifold to the choke. Some engines have another tube which parallels this tube. Connect it to the tube coming from the right-rear underside of the carburetor base. This tube provides filtered air to the choke heat tube at the exhaust manifold. easy to cross-thread it. Just make sure the tube is bent so it has a "straight shot" at the choke. Not all engines, but some have another tube which parallels this tube. Connect it to the tube coming from the right-rear underside of the carburetor base. This tube provides filtered air to the choke heat tube.

Carburetor Linkage—You will have either a single rod connecting the carburetor to the accelerator-pedal lever at the firewall or it will go to a bellcrank with mounts on the intake manifold and then a second rod connects the bellcrank to the accelerator. In addition, there is a rod which parallels this one which connects the bellcrank to the firewall. If you are puzzled by this arrangement, it's to cancel the effect of engine roll on the position of the throttle. The latest linkage used is the cable type. When a cable is used in conjunction with an automatic transmission, you'll also have a kick-down rod to attach to the throttle-plate lever.

For attaching the simple lever-type linkage, position the retaining clip over the throttle-plate-lever hole first, then insert the bent end of the rod in the hole. You can now rotate the clip over the rod. For the bellcrank type, mount the bellcrank to the intake manifold and then the rod to the carburetor unless you removed the bellcrank from the engine compartment rather than just rotating it up out of the way. If this is the case, you'll have to hook up the other two rods, also.

The final customer is the cable type. Just insert the cable through the cable bracket attached to the intake manifold. Push it far enough so the end of the cable conduit clips into the bracket. There'll be a ball-pivot extending from the throttle-plate lever. Push the end of the cable over it until it clips into place. If you have an automatic transmission, hook the kick-down rod by slipping it over a pin on the throttle-plate lever and retain it with the clip you should've left on the carburetor. The last thing to do with the throttle linkage is to install the return springs. Earlier rod types have a bracket at the right-rear carburetor-mounting stud. Attach the spring between this bracket and the zig-zag in the throttle rod. The bellcrank type has its return spring on the bellcrank itself. It should still be there because you didn't have to remove it. The cable type has one return spring as part of the cable assembly and also uses an additional spring which hooks between the throttle-plate lever and a bracket on the left-front carburetor-mounting stud. Install the spring between these two points.

Ignition Coil—Mount the ignition coil to the intake manifold or the front of the left cylinder head. If your coil was originally mounted to the manifold, relocate it to the head, or to the A/C bracket in a vertical position. The reason for this, particularly in hot climates, is that under-hood temperatures reduce a coil's output which usually results in coil failure. You may have to modify the coil-mounting bracket slightly so it will mount in these other locations, but the extra effort now will pay off later, particularly if you've had recurring ignition problems.

Wiring Harness—Locate the engine-wiring harness along the inboard side of the left valve cover. Retain the harness in the clips you've installed under the valve-cover bolts and connect the leads to the oil-pressure and water-temperature sending units and to the ignition coil. If your car has A/C, there'll be another wire separate from the main harness which is routed along with the main harness. This lead operates the A/C compressor clutch.

Fuel Filter and Line—The biggest variation you'll find in the fuel-pump-to-carburetor plumbing is the type of filter your engine is equipped with and its location. The latest and commonest type threads into the carburetor float bowl. Another is the in-line type followed by the very early cartridge type. Regardless of which one your engine uses, install a new filter and complete the line routing.

Set Number-1 Cylinder on TDC—A distributor should be installed so it is in position to fire the number-1 cylinder. This also requires that the piston in the number-1 cylinder be set on the TDC of its *power stroke*. Remember, the number-1 TDC mark on the crankshaft damper passes the timing pointer *twice* for each complete cycle, once between the exhaust and intake strokes and once between the compression and power strokes. The later is the one you want.

There are several ways of determining when a piston is at the top of its power stroke, but I'm just going to discuss the easy method. It doesn't require using much more than your common sense. Start by removing the number-1 spark plug—that's why I suggested leaving it loose earlier. Put your thumb over the hole, then crank the engine over. You should feel the compression lift your thumb once every two revolutions as the piston approaches the top of its power stroke. Now that you have a feel for it, bump your engine slowly while keeping your eye on the damper. As you feel your thumb beginning to lift, the timing mark should be approaching the pointer. When it's in sight, stop. Bring the TDC mark within 10° of the pointer and you're ready to install the distributor. You can now reinstall the spark plug and torque it 15—20 ft. lbs.

Installing the Distributor—With the engine timed, you can now install the distributor. To be able to "time" the distributor with the engine, you need a way to determine the position the distributor rotor should be in to fire the cylinder 1. Use the distributor cap as a reference to mark the distributor housing for locating the rotor.

Hook up your carburetor linkage and adjust it for wide-open throttle. Don't forget the return spring that goes between the tab (arrow) and the 90°-bend in the link. Also check the choke-plate operation. It should be free.

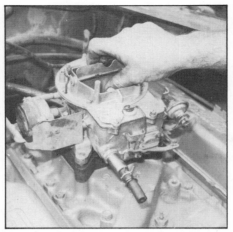

An equalizer-type accelerator linkage connects to the firewall (arrow). Equalizer-to-carburetor link attaches directly above the front of the return spring.

Mark the side of the housing with a grease pencil in line with distributor-cap socket 1, then remove the cap. This will enable you to locate the rotor in the position it should be in to fire the number-one spark plug at the right time.

Before installing the distributor, double check to make sure there's an "O-ring" installed on the base of the distributor housing, then smear some oil on it. When installing the distributor assembly, keep

Route the engine wiring harness along the inside of the left valve cover. Retain it with clips under the valve-cover bolts. Mount the coil vertically and make all the harness connections—to the oil-pressure and water-temperature senders and the coil. If you have A/C, you'll have to mount the coil to the front of the compressor bracket.

First step in the distributor installation process—check number-1 cylinder for TDC by rotating pushrods with crank set on TDC. If pushrods are tight rotate crank 180° and recheck pushrods.

the vacuum-advance can pointed forward and start with the rotor pointed rearward. The photo shows the outer tip of the rotor is approximately one inch away from the mark. Here's why: as the distributor-drive gear engages the camshaft gear, the rotor will rotate clockwise bringing the rotor in line with the mark, providing you keep the vacuum-advance can pointing straight ahead. If the rotor doesn't line up with your mark on the housing pull the distributor out far enough so you can turn the rotor and shaft and try it again. A little care here will ensure that correct timing can be achieved without positioning the distributor housing at an awkward angle.

If you have a high-performance 289 which uses centrifugal advance only, meaning it doesn't have a vacuum can, position the distributor housing so the number-one cylinder mark is at the rear and slightly to the left. Also, the distributor-cap retaining clip should be to the left of the water-temperature-sending unit. When it's swung down toward the intake manifold, it should clear the sending unit.

Oil-Pump Drive Shaft—One complication you may run into is installing the distributor. In fact, I can be almost certain of it unless everything goes right for you. Because the distributor shaft turns the oil pump too, the shaft must engage the cam gear as well as the oil-pump drive shaft. This being the case, you'll probably find that the distributor won't install completely into the engine on your first attempt. It would be pure chance that the oil-pump drive shaft will line up with the mating end of the distributor shaft. As a result, you'll have to bump your engine over until the distributor engages the drive shaft so the distributor can drop into place with some assistance. Return cylinder 1 to TDC and check to see if the rotor and the mark on the distributor housing line up. If not, try again by lifting it out carefully and indexing the shaft one tooth in the opposite direction it's out. You'll probably have to go through the oil-pump drive shaft engagement procedure again.

Now that your distributor is in place, you can come fairly close to correct timing before the engine is run by *statically* timing it. To do this, turn the crank so the pointer and crank damper indicate the correct distributor advance—10° BTDC, for example. Install the distributor hold-down clamp and bolt loosely, and then rotate the distributor housing clockwise, or against the rotation of the rotor until the points *just* begin to open. The ignition coil releases its voltage the instant the points *break* contact, firing the spark plug. As soon as you have the distributor where you want it, secure the hold-down

 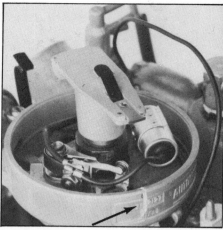

Install your distributor after confirming cylinder 1 is on TDC of its firing stroke. As the distributor gear begins to engage the cam gear, you'll probably have to bump the engine over slightly before it engages the oil-pump drive shaft and drops into place. After your distributor is fully installed, set your engine back on TDC of its firing stroke. The rotor should line up with the number-1 cylinder mark (arrow) with the vacuum cam pointing forward. Clamp the distributor housing in this position.

Install a set of new spark plugs after you've gapped them. Route the ignition wires according to your engine's firing order beginning with the number one cylinder: 1-5-4-2-6-3-7-8 for all 221, 260, 289 and 302 CID engines and 1-3-7-2-6-5-4-3 for the 351W. Using valve-cover clips to hold the wires in place, route them neatly according to their lengths. Cylinder 7 and 8 wires for all but the 351W should be separated in the left valve-cover clip (arrows). Separate the 3 and 4 wires in a similar manner for a 351W.

Bypass hose gets replaced regardless of how old it is. Because these hoses are usually too long, trim yours so it doesn't pinch closed when installed.

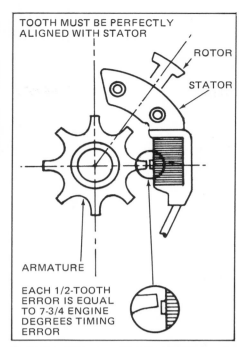

Statically time a solid-state distributor by rotating the housing so one armature vane lines up perfectly with the magnetic pickup as shown in the circle. Rotor should also be located midway between the two magnetic-pickup rivets. Drawing courtesy Ford.

clamp. Later, when you get your engine started, check the timing with a timing light and readjust it if need be.

If your distributor is a '74–'78 breakerless type, a similar method works. However, rather than having the points to go by, align one of the armature *spokes* as shown in the sketch. Looking down on the distributor, the rotor should be between the two rivets which secure the magnetic pickup when it is in the number-1 firing position.

Thermostat Bypass Hose—Regardless of age, I always replace the thermostat bypass hose for no other reason than it's so much easier to do now than a month or two years later. New bypass hoses are longer than they need to be, consequently, you'll have to trim the excess length off the ends. To judge how much you'll have to cut off, set the hose alongside the water-pump and intake-manifold inlets and mark it. Cut the hose and recheck it for fit, then install it. If the hose isn't

Bracket has to go back on A/C compressor before it can be mounted to the engine. With compressor in place, install power-steering pump followed by compressor's fixed-idler pulley and bracket. To line up this pulley, make sure you have spacer (arrow) between idler bracket and power-steering-pump bracket. Install top compressor bracket, being careful to install bolts under and behind the bracket. Finish the job by installing compressor strut which supports it from the manifold.

trimmed, it will kink and restrict proper coolant circulation through the engine when the thermostat is closed.

ACCESSORY DRIVE INSTALLATION

The degree of difficulty of installing your accessory drive can vary from a five-minute job to the toughest part of the engine installation. I'll treat the installation like your car is loaded and you can simply delete those sections that don't apply to your installation.

A/C Compressor and Power-Steering Pump—Rather than treating the A/C compressor and power-steering pump separately, I'll have to deal with them as a team because they use common mountings—not totally, but enough so one can't be mounted without the other. This is only true when you have both accessories.

Your A/C compressor should still be hanging where you left it when removing your engine—on the fender apron. While it's still there, install the mounting bracket to the bottom of the compressor, the big bracket. Make sure you use the right bolts. They are the short 3/8-16 bolts with the integral serrated-face washers. Ford terms these *UBS bolts,* meaning uniform bearing strength. Make sure you use the correct length bolts, install them carefully to check for "bottoming." The very short ones are used to mount the adjustable-idler bracket to the compressor bracket. Torque these bolts 20–30 ft. lbs. You can mount the compressor to the left cylinder head. Torque the bolts 23–28 ft. lbs. When initially mounting the compressor you need an accomplice to start the bolts while you support the compressor. An A/C compressor with its clutch is a little too much for one guy to handle alone.

Next customer is the power-steering pump and bracket assembly. It mounts to the A/C compressor bracket and to the water pump. The bolts at the water pump not only support the power-steering pump, they also mount the water pump. Torque the bolts at the water pump 15–20 ft. lbs. and the one at the compressor bracket 23–28 ft. lbs. One thing to watch out for at this point is what looks like a thick washer on one of the water-pump bolts. I say "looks like" because it's actually a spacer and not a washer and it goes on the bottom power-steering-pump-bracket-to-water-pump mounting bolt. Leave it loose, it mounts even more!

A/C-Compressor Idler-Pulley Bracket—Turn your attention back to the A/C compressor. Install the bracket which mounts the adjustable pulley. Three UBS bolts mount this bracket, two on top of the compressor and one which is effectively hidden behind and under the bracket. Don't miss this last one. Also, install the clip for the compressor's magnetic-clutch lead and the A/C specification tag under the top-rear mounting bolt. Torque the bolts 20–30 ft. lbs. and make certain the bolts are short enough so they pull down on the bracket rather than bottoming in the threaded holes in the compressor housing. If it's not already on the bracket, install the idler pulley, but don't tighten it yet.

Fixed Idler Pulley—Not all air-conditioning equipped engines use a *back-side* idler pulley. This fixed pulley stabilizes the belt to prevent it from *whipping.* If yours has one, you can easily recognize it by its bracket—it looks like a three-legged octopus. This is because of the different planes in which the bracket mounts as well as the spread of the mounting holes. The top leg mounts to the adjustable-idler-pully bracket and the lower-inboard leg to the water-pump bolt with the spacer. Make sure this spacer goes **immediately behind** the idler bracket, otherwise the pulley won't line up with the other pulleys and will result in excessive belt wear or a belt which is constantly jumping its pulleys.

A/C Compressor Wiring—To finish the accessory installation job on the left side of your engine, connect the compressor's magnetic-clutch lead. Do it now because it's easy to forget. You'll wonder why your air conditioning won't work the first time you try to use it.

Air-Injection Pump and Alternator/Generator—Just like the A/C compressor and power-steering pump, if your engine is equipped with an alternator and an air pump, they'll use common brackets. However, unlike the A/C compressor and power-steering pump, the position of the alternator and air pump can be different, depending on the year and model of your vehicle. Because you should already have your alternator mounted, you'll just have to install the air-injection pump. If yours is the type which mounts above the alternator, it'll be a little easier just because of the easier access. Regardless of the relative positions of the two units, the air pump mounts off the alternator pivot and is supported on the opposite side by an adjustable arm similar to that used for the alternator. Mount the air pump, but don't tighten it up yet. Route the air-pump hoses and mount the bypass valve if there are provisions for doing do. Bypass valves used with air pumps mounted above the alternator are usually supported by their hoses. The others mount from a tab supported by the top thermostat housing bolt. One hose runs from the air pump to the bypass valve, but one or two go from the bypass valve to the check valve/s. The type using one valve is internally manifolded and locates its check valve at the back of the engine whereas the others are externally manifolded along the side of each cylinder head. Before you install the air-pump hoses, inspect and replace them if cracks appear when you squeeze them—in this case, they're about ready to fail.

Spark Plug and Coil Leads—Turn your attention back to the ignition system. Make sure the rotor is fully installed—all the way down on the distributor shaft—and install the cap and secure it with the two spring clips. If they're not already there, install the little plastic ignition-wire-retaining clips. Starting with socket 1 on the distributor cap, route the wires in the engine firing order as you go around the cap counter-clockwise, point the wires in the general direction of the valve-cover clip they will have to be routed through. Remember, the firing order for all small blocks *except* the 351 is 1-5-4-2-6-3-7-8. The 351 order is 1-3-7-2-6-5-4-8. The order for the wires in the right valve-cover clip is 1-2-3-4.

Because of the difference in firing orders between the 351 and the rest of the engines, the order for the 351 left valve-cover clip is 5-6-7-8. The order for the rest is 7-5-6-8. The reason for the odd order for the left-hand valve-cover clip is to separate the wires running to cylinders 7 and 8 which fire right after one another, for all small blocks except for the 351 and 6 and 5 for the 351. This prevents a possible misfire in the second cylinder caused by *induced current* in the second wire—number 5 in the 351 and number 8 in all the others. An induced current occurs when two wires are close and parallel for some distance. Current flowing in one of the wires automatically induces current to flow in the other wire, enough to cause a weak spark and a misfire.

Spark-plug wires have different lengths. Consequently, the easiest way to install new wires, if you're doing so, is to install the old distributor cap and wires. Now, all you have to do is replace each wire one-at-a-time by comparing the lengths and duplicating the routings. However, if you've discarded your old wires you'll have to do a trial routing. Install the wires loosely at the distributor cap. They are difficult to get off if they are fully installed. After you are satisfied with the routing, push down on the molded cap at the end of each wire to seat it in its distributor-cap socket. It's not easy because, as the cap goes on, air has to be forced out. As for the spark-plug end, push until you feel the connector click into place, then give it a little tug to see if it's on all the way. Pull on the molded sleeve, not on the wire. Don't forget the coil lead. It's easy to install once the plug wires are in place.

Heater Hoses—Install new heater hoses as you would a radiator hose. If they are over two-years old, replace them and use the old hoses to determine the length of the new ones. The heater-inlet hose is routed from the intake manifold fitting to the *bottom* heater-core tube. The

Some heater hoses are used to heat the automatic choke. If yours is this type, route the hose from the intake manifold to the bottom heater-core inlet between the choke and the adjacent clip.

heater-outlet hose is routed from the *top* heater-core tube to the water-pump inlet. Route the heater-outlet hose through the automatic-choke retainer on models that provide one.

Miscellaneous Hoses—The last major job is to route the vacuum hoses. Again, the complexity of this depends on the type and number of accessories your car is equipped with in addition to the emissions devices used. Here's where your labeling of the hoses and photos you may have taken will be invaluable. Before you begin installing the hoses, check them for signs of cracking. If some have begun to crack, replace all of them. A cracked vacuum hose can be terribly hard to find and it really complicates getting your engine to run right.

Depending on your particular vehicle, possible vacuum hoses you may have from various vacuum taps at the back of the intake manifold are for: power-brake booster, automatic-transmission throttle valve, heater and air conditioning vacuum motors, the distributor-advance diaphragm and emissions devices such as an EGR valve (exhaust gas recirculation) routed through a PVS (ported vacuum switch) in the thermostat housing. The PVS meters engine vacuum to the distributor and EGR valve in some instances, but not all. The scope and type of emissions equipment and the vacuum circuitry is beyond the scope of this book, so the correct routing of your system depends on how well you labeled or photographed your engine during the removal process.

Engine Ground Strap and Automatic-Transmission Filler Tube—Because of their common location and difficult access, install the engine ground strap and the transmission filler tube to the back of the right cylinder head. Having double-jointed fingers is helpful. When attaching the ground strap, make sure you install a *star washer between* the ground-strap connector and the cylinder head. The washer ensures a proper ground between the connector and the cylinder head, particularly in geographic locations where corrosion is a problem.

Water-Pump Pulley and Fan—In preparation for installing the radiator, install the water-pump pulley and the fan. To make the job easier, locate the pulley on the water-pump shaft and line up the pulley holes with those in the water-pump flange. If you have a fixed fan, loosely assemble the fan, spacer, the four mounting bolts and lock washers and locate this assembly on the water-pump shaft and start the bolts. Tighten the bolts in a criss-cross pattern to 10–15 ft. lbs. You'll have to grip the fan to prevent it from turning while tightening the bolts.

If your fan is the flex-blade type, inspect the blades for cracks. Because of the constant flexing, the root of the blade eventually becomes fatigued to the point of cracking and breaking off. As a result, this type of fan has been known to throw a blade when the engine is running, so remember, **never stand in line with the fan when the engine is running,** particularly when the engine is operating at higher than idle RPM.

When installing a clutch-drive fan,

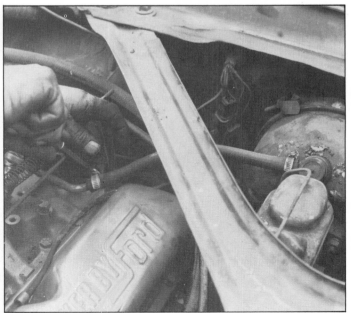 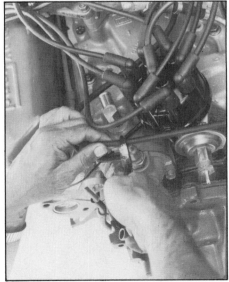

Fitting and installing some of the many possible vacuum hoses. Large hose supplies vacuum to the power-brake booster. When fitting new hoses lay them out on the engine and cut them to length. One being cut connects distributor vacuum-control valve (PVS valve) to intake-manifold fitting behind carburetor.

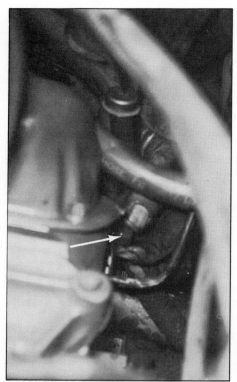

Don't forget this bolt. It secures the engine ground and automatic-transmission filler tube to the back of the right cylinder head—and it's a tight area. Your engine ground strap may attach differently. Just don't forget it.

Match water-pump-pulley holes to those in water-pump-shaft flange before trying to install fan and spacer. Run fan bolts down and torque them 10—15 ft.lbs. Now install the accessory-drive belts. Adjust them to a 1/2-in. deflection and secure the tensioners.

there's no spacer, but the job may still be a little more difficult because of the way the bolts are located behind the fan and clutch-drive assembly. The fan is not fixed, so you'll have to wait until the drive belts are installed before you can torque the mounting bolts. So for now, tighten them snugly as you hold the water-pump pulley. Complete the job after you've installed and adjusted the drive belts.

Accessory-Drive Belts—Depending on the accessories, your engine will be equipped with from one to three belts. Again, replace them if they are more than two years old, but inspect them regardless of their age. If you have more than one drive belt, install them in the proper order back to front.

It's very important that the belts be adjusted to their proper tension. If one is too loose, it will slip in the pulley of the accessory it drives. If it happens to be the alternator/generator drive belt, you'll be plagued by a battery that seems to run down for no reason at all. Conversely, if a belt is too tight, it will overload all the bearings in the pulleys and shafts it drives. This is particularly critical in the case of a water pump. They don't take kindly to being overloaded. It shortens a water pump's bearing life drastically.

As for how tight a drive belt should be, Ford specifies 140 lbs. tension in a new belt and 100 lbs. in a used one. A used belt is defined as one that has been used for 10 minutes or more. Knowing the correct tension and measuring it are two different matters. There is a belt-tension checking tool, but it rarely finds its way into even the most complete tool chests. If you happen to have one, by all means use it. I recommend using the *deflection method* for checking belt tension. Push firmly on the belt (about 10 lbs. pressure) and measure how much it deflects. A properly adjusted belt should deflect approximately 1/2-in. in the middle of an unsupported length of 14—18 in. Any more means the belt is too loose. Less means it's too tight. For new belts, the deflection can be slightly less, but after it's run for a while, no less than 10 minutes, recheck it. The only equipment you'll need for checking a belt's deflection is a straight edge and a ruler or something indicating 1/2 in.

Depending on the year of your car and how it's equipped, there are three possible ways to adjust a drive belt. The most common, and the one you will have regardless of the year and how your car is equipped, is the type requiring that the accessory be rotated. All alternators and generators and some power-steering pumps use this method. To adjust a belt, rotate the accessory with a pry bar or a large screwdriver so the belt will be held in tension while you tighten the adjusting-arm bolt to maintain this tension. When doing this, the pivot bolt and adjusting-arm bolt should be loose. Just make sure you don't pry against something that can be damaged, such as the reservoir of a power-steering pump. Torque the bolts 25—35 ft. lbs. on the power-steering pump and tighten the alternator/generator bolts securely.

The second method is the A/C compressor drive belt adjuster. Rather than rotating the accessory, an adjustable idler pulley rotates against the slack-side of the drive belt. Its adjustment is convenient because the 1/2-in.-square hole in the pulley bracket accepts a 1/2-in.-drive breaker bar. The guy that came up with this good idea must have personally busted his knuckles a few times using the previous

Radiator being installed *after* the fan shroud is laid back over fan and front of engine. With radiator in place, shroud can be installed to it.

A dab of white paint on the damper pulley on the proper advance mark shows up well under the strobe light. Disconnect and plug any distributor vacuum lines when setting and checking your timing.

Don't forget the rubber seal when installing your air cleaner, otherwise dirt will be inhaled by your engine. Start with a new air-filter element, too.

method of belt adjustment. To adjust belt tension, rotate the pulley bracket in the clockwise direction with your 1/2-in. breaker bar and secure the bolt in the slotted hole. Check belt tension, then secure the pivot bolt when you've reached the correct adjustment.

The last method of adjusting belt tension is the ultimate when it comes to the present state-of-the-art. It's used for adjusting many of the mid-1970 and later power-steering pump drive belts. Rather than rotating the pump to adjust belt tension, the pump uses two brackets that slide on each other. To adjust a belt using this type, loosen the three vertical retaining nuts, clamping the two brackets together, then rotate the nut on the adjusting stud which extends through the upper bracket. Turn the nut clockwise to tighten the belt. Be careful when doing this, it's easy to over-tighten a belt because little effort is required as opposed to the other methods. When you've reached the correct belt tension, torque the three clamping nuts 30—40 ft. lbs.

Radiator and Fan Shroud—Before installing your radiator, position the fan shroud—if you have one—over the fan. If you don't do this, you'll either have to remove the fan or radiator again to install the shroud. Carefully lower the radiator into place and set it on the rubber pads if you have the type that is mounted in rubber, or start the mounting bolts through the radiator flange if yours mounts solid. For rubber-mounted radiators, install the upper radiator bracket. On solid-mounted radiators, install and tighten the remaining mounting bolts. Now you can slip out the cardboard you used for protecting your knuckles and the radiator core. If your car is equipped with an automatic transmission, remove the hose you used to connect the transmission cooler lines and connect the lines to the radiator fittings at the bottom or side tank for down or cross-flow radiators, respectively. Use a 5/8-in. tube wrench to tighten the nuts to avoid rounding them off and don't over-tighten them—12 ft. lbs. is maximum. You won't be able to use a torque wrench here, but you should've developed a "feel" for torque by now. Install the upper and lower radiator hoses.

Engine Run-In Preparation—Prior to the initial starting and running of your newly rebuilt engine, what you do now will ensure a long life for your engine. The first 30 minutes of running are the most critical. The rules are, once the engine is first started, it should be kept running for 30 minutes, it must be well lubricated and cooled. Lubrication is particularly critical for camshaft break-in. If the cam lobes aren't drenched in oil during the initial run-in, one or more lobes can be wiped off. This means tearing your engine down again, so I'm certain you don't want this situation. As for cooling, it is obvious that an engine needs to be cooled when it's running and your car will be stationary and not be getting full air-flow through the radiator. In addition, a new engine being run-in will develop more heat than a broken-in engine. The reason for this is friction. Tighter clearances will exist between all moving parts. As your engine is being run-in, you'll notice RPM will increase as it runs with a given throttle setting, meaning the engine is loosening up.

With these three points in mind, let's take a look at how you can make them all "happen."

Cranking an engine accomplishes two things. The first is to confirm that your engine will build oil pressure within 15 seconds: 35—40 psi with a standard pump and up to 60 psi with a high-volume pump. If you have an idiot light, all you'll

151

be able to do is check for the idiot light to go out indicating pressure. The second thing you're after is to get fuel to the carburetor and fill its float bowl/s. This will ensure that once the engine has started, it will continue to run. Your engine will spin faster and there will be less load on your starter if you remove the spark plugs, then replace them for starting.

As for the cooling aspect of the engine run-in, to make sure your engine is being well cooled during all of the first 30 minutes, put your garden hose into service. Fill the radiator with water, then open the drain and adjust the water flow from the hose to the radiator to match what is coming out of the drain. This ensures a constant source of cool water to the engine.

Engine Run-In—Replace the plugs if you removed them and make sure the fast-idle cam on the carburetor is in its fast-idle position. Start your engine. **Avoid letting your engine idle at slow speeds.** It should idle for 30 minutes between 2000 and 2500 RPM. A new engine idled at low RPM could ruin its camshaft due to insufficient lubrication.

With the water running through the radiator, start your engine. Don't be frightened by the initial puff of blue smoke and the clattering lifters. They should both remedy themselves after the first few minutes of running unless you have the solid-liftered HP289 or Boss 302. Noisy lifters is one of the joys of these engines—if you like that. Adjust the idle if it's not right and keep your eye on the radiator. When the thermostat opens, the water level will go down as water is drawn into the engine, consequently, you'll have to add more before the water flow in and out stabilizes again.

If you have an automatic transmission, the torque converter will fill with fluid during engine run-in, lowering the fluid level in the transmission. This will be indicated on the dip stick, so add fluid accordingly.

After running your engine for the 30 minutes, shut it off and connect your timing light to the number-1 plug. Using a box-end wrench, loosen the distributor cap and disconnect the vacuum-advance hose and plug its open end. Restart the engine, but idle the engine down this time. Point the timing light at the crank damper and pointer. Adjust the distributor for proper advance and lock the clamp. Turning the distributor clockwise advances the timing; counterclockwise retards it. Shut the engine off and reconnect the vacuum hose. BEWARE: When working around the front of your engine while it is running, the fan is just waiting to grab a shirt tail, sleeve, cord or whatever to do serious bodily harm. The same thing goes for accessory drive belts and pulleys. **Be careful. Stay out of line with the fan.**

Post Run-In Checks—After initial run-in, shut the garden hose off and let the water drain out of the radiator. When it stops draining, shut the drain cock in preparation for refilling the radiator with permanent coolant. Check your engine all over for fluid leaks; gas, oil and water. If you spot any, remedy them.

Because gaskets and hoses *creep*, or *relax* when they are loaded and heated, you should go over the entire engine and retighten a few things. High on the list are the intake-manifold bolts, heater-hose and radiator hose clamps. While you are at it, check the exhaust-manifold bolts too.

Engine Coolant (should be anti-freeze)—Regardless of the climate your car will be operated in, water should not be used as the sole coolant. If you use it as the main coolant, rust inhibitor must be added to prevent the interior of your engine from being damaged. I believe the best practice is to use anti-freeze so your coolant will have at least 0°F ($-18°C$) capability. This provides corrosion protection for your engine and raises the temperature at which the coolant can operate before boiling.

Fill your radiator until it won't accept any more, then start the engine and wait for the thermostat to open. Have some anti-freeze handy. The engine will begin to purge itself of air, so you'll have to keep an eye on the coolant-level in the radiator. Don't forget the heater. To fill it, put the heater control on *heat* so it opens the water valve if your heating system is so equipped. Coolant will now be able to circulate through the heater core and hoses. When the coolant ceases to go down and is free of bubbles, cap the radiator. If you have the coolant-recovery type system, fill the recovery reservoir to the indicated *hot level*. Install a new radiator cap with the correct pressure rating.

Air Cleaner—You should now be ready to install the engine's crown, the air cleaner. Make sure there's a seal on the carburetor before installing the lower part of the air cleaner. If this area is not sealed, dust and dirt will enter the carburetor rather than being filtered out of the air. If your engine is equipped with an exhaust-manifold heat duct or fresh air duct, connect them. Install a new filter and put the top on the filter. Secure it with the wing nut. but make sure there's a seal under the nut for the same reason as the seal between the air filter and the carburetor. Connect any hoses or wires your air cleaner assembly may require, such as the PCV hose.

Hood—You need to rustle up some help to install the hood. With the hood held in place, run the attaching bolts in just short of snug and adjust the hood to the reference marks or drill holes you made

After completing your preliminary engine running, drain your cooling system and recharge it with a good mix of anti-freeze. In addition to its anti-corrosive qualities it will also prolong water-pump-seal life.

prior to removal, and tighten the bolts. Close the hood to check its adjustment. It should fit the same as it did before removing it.

Trial Run—Now, with the satisfaction of having rebuilt your own engine, it's time to take it on its maiden voyage. Before you go tooling out of the driveway, make sure you've collected all your tools from the engine compartment. Your wrenches aren't going to do you any good distributed along the roadside. Take an inspection trip around your car to make sure you don't end up flattening your creeper, or whatever may have been left under your car in your haste to "see how she's gonna run."

When taking your engine for its initial run it's not going to hurt it if you take it up to highway speed, just avoid accelerating hard. Also, vary your speed. Drive just long enough to get your engine up to normal operating temperature and stabilized. Keep alert for any ominous sounds, then get back home for a look-see under the hood. Check fluid levels. Look for any leaks that may have developed and generally scrutinize the engine compartment. After everything appears to be OK,

Installing the hood by lining up the hinges to the reference marks. Start by installing the rear bolts first. Assuming your hood fit before you removed it, it should fit after it is reinstalled.

retorque the intake- and exhaust-manifold bolts. You can now admire *your engine* unless...if your engine is equipped with solid lifters, you have one more duty.

Mechanical-Lifter Hot-Lash Adjustment—If you have the solid-liftered HP 289, Boss 302, or have installed a solid-lifter cam, you'll have to readjust the valves in their *hot condition*. This is because the growth of the valve-train components as they heat up changes the valve lash—and does so inconsistently. Consequently valve clearances must be reset with the valve train at its normal operating temperature. The 0.022-in. cold setting you made was an approximation, but not a guarantee of creating the specified hot-lash of 0.018 in. for all the valves, so readjust them.

To adjust valves at their operating temperature they must be operating. This is hot and messy. To make it easier, acquire some *rocker-arm clips* and a 0.017–0.020-in. *go–no-go feeler gauge*. The rocker-arm clips fit over the pushrod end of the rocker arms and prevent oil from being thrown all over the side of your clean, newly rebuilt engine, your engine compartment and you. Besides being messy, the oil is hot and will smoke when it splashes on the hot exhaust manifolds. *Rocker-arm clips*, which clip to the rocker arms to divert the oil back into the top of the cylinder head, are available at most automotive parts stores and hot-rod shops. Also, go–no-go, or step-type feeler gauges are much easier to use than the more common single-thickness type, especially when you're adjusting the valves while they are operating.

Before adjusting your valves you'll first have to warm up your engine to its normal operating temperature. A short drive should accomplish this. After the engine has been warmed up shut it off and quickly remove the valve covers, being careful not to damage the gaskets. Tie the ignition wires up out of the way and install the clips on the rocker arms. Have your feeler gauge and a 5/8-in. socket with a non-ratcheting handle ready, then restart your engine. Work from one end of the head to the other. If you're using the go–no-go feeler gauge, it should slide freely between the valve tip and the rocker arm on the 0.017-in. thickness, but the 0.020-in. thickness shouldn't fit at all when you've obtained the correct 0.018-in. clearance. Remember, the valves are going to be opening and closing, so the feeler gauge will be intermittently trapped between the valve and the rocker arm between the time the lifter is not on the cam-lobe base circle. After going through all your valves, shut your engine off and remove the valve clips.

A tip is in order at this point. If you've lived with an engine that's equipped with a solid-lifter cam for any length of time you'll know how often the vlaves must be adjusted to maintain proper lash—about every time your engine is due for a tune-up, or every 10,000 miles. This can become a nuisance after a while. The reason for this constant adjustment is the valve-adjusting nuts *back off* gradually. So to extend the amount of time between adjustments install a *jam nut* like the Boss 302 on top of each rocker-arm adjusting nut. This will prevent the valve lash from changing due to the adjusting nut turning. You can use a standard 3/8-24 jam nut or a sheet-metal Pal-nut like those used for locking some connecting-rod nuts. Your

If your engine is equipped with a mechanical cam you'll need eight rocker-arm clips like these from Mr. Gasket®. They'll reduce the amount of hot oil that gets splashed on you and your engine compartment when you're doing hot-lash adjustments.

Ford parts man has them under part number 45218-S8. These are for locking the HP289 connecting-rod nuts. Install these snugly, not tight because they can't take much torque. Be careful not to turn the adjusting nuts. Put a wrench on the adjusting nuts to prevent this from happening.

Now you can reinstall the valve covers, but first wipe any oil off the valve-cover gasket surface. Reattach the ignition wire clips to the tabs on the valve covers. This should pretty well button the engine up. If you didn't use rocker-arm clips you might want to take a trip to the local car wash to clean your freshly oiled compartment.

While putting the first 200 miles on your engine, constantly check the fluid levels—particularly oil—and look for any leaks. Keep checking things in the engine compartment in general. Correct any problems that might arise. Don't get excited if your engine uses a little oil, particularly if you've used chrome piston rings. Just keep the oil level up and change the oil and filter after the first 500 miles. Make the first miles easy ones and avoid operating your vehicle at sustained or steady speeds. Keep varying the speed. Avoid accelerating hard. Follow these rules and it will pay off in a longer lasting engine. After you've put 200–300 miles on your engine, take your car or truck to a tune-up shop—one that specializes only in tune-ups—and have your engine expertly tuned. Now, your engine will not only be well broken in, it will perform at its peak efficiency. For more information on engine tuning see the next chapter.

Tuneup 10

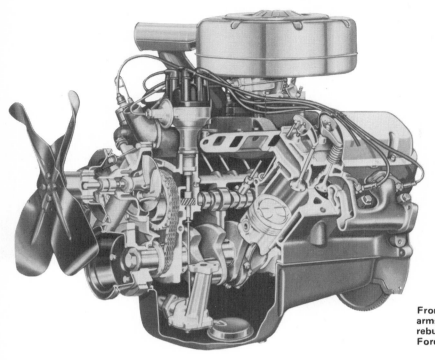

Front quarter section of a '65 289 using pushrod-guided rocker arms. Getting the most from the internal parts of your newly rebuilt engine requires that it be properly tuned. Photo courtesy Ford.

After putting a few hundred miles on your newly rebuilt engine you should have it professionally tuned so it will provide you with peak performance. Even with the most complete tune-up set you're not going to be able to do much more than set distributor dwell, initial timing, check point resistance, total advance and set your carburetor's automatic choke and idle settings. This is particularly true with the later engines using electronic ignitions and complex emissions systems. Remember, you should not only be tuning for performance but for minimum emissions. The biggest polluters on the roads are vehicles which have untuned or improperly tuned engines. Typically, these cars do not perform as well as they could and they do not run as economically as they should.

PERFORMANCE AND EMISSIONS TUNING

The cost of a complete professional tune-up should range from $25 to $75 depending on the work required. The cost to you should be at the low end of the scale if you've done a proper job of rebuilding your engine.

The type of tune-up equipment to tune an engine depends on how it's built and equipped. For instance, if you rebuilt your engine to original specifications an electronic engine analyzer can be used. This will ensure your engine will perform *at least* as well as when it was new. However, if you modified your engine such as installing a different camshaft, changing the carburetion system, ignition system or exhaust system your engine must be recalibrated for it to obtain maximum performance. The only way of doing this is to tune it while its performance can be "seen" on a *chassis dynamometer*. The rear wheels of the car or truck drive *rolls* that are set flush in the garage floor. These rolls are adjusted to simulate the load the vehicle would "see" on the road. Therefore, the chassis dynamometer permits an engine's functions to be monitored while it is operating under simulated driving conditions. In addition, the dynamometer can be used to determine the engine's power *at the drive wheels* which means an engine can be tuned for maximum power *and* mileage and emissions. Beware of tuning for full power at full throttle only. This will compromise the more useful areas in the "performance curve." Tuned correctly, dividends in terms of power, fuel mileage and emissions will more than pay for the initial tune-up investment and ensure maximum pleasure and benefit from your newly rebuilt small-block Ford.

Having your engine professionally tuned is the best way of getting the most out of your newly rebuilt engine. If your engine was built to original factory specifications, an electronic engine analyzer like the one pictured is sufficient to do the job.

Checking horsepower where it counts, at the rear wheels. Though increasing in use, not all tuneup shops have chassis dynamometers. Even fewer have dynamometers capable of checking full horsepower, so check first if you want full horsepower readings.

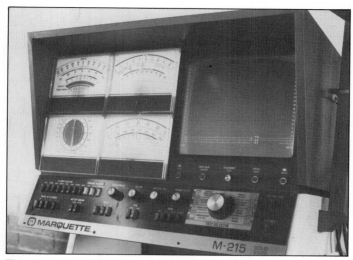 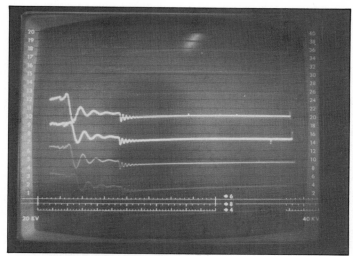

This electronic engine analyzer is capable of monitoring air/fuel ratio from the carburetor and distributor dwell at various engine RPM. Oscilliscope also monitors performance of each spark plug as shown in the scan.

Exhaust gases being analyzed. Comparing hydrocarbon and carbon-monoxide outputs is an actual situation method of determining whether an engine's fuel is being used efficiently during the tuning process. This device is valuable both when tuning for economy and in for tuning for reduced emission levels—an increasingly important consideration.

OFFICIAL SHOP MANUALS
For additional information concerning your car or truck you should obtain the official Ford shop manual. Single-volume manuals covering specific vehicles were printed for trucks prior to 1966 and for cars prior to 1970. These single-volume manuals were superseded by five-volume manuals which cover *almost* all car or truck lines. They are grouped according to the vehicle system—chassis, engine, electrical and body. The fifth volume covers general maintenance and lubrication. These manuals are available from Helm. To order, or for additional information, contact Helm, Inc., P.O. Box 07150, Detroit, Michigan 48207. Telephone: (313) 865-5000. Make sure you include the model and year of your vehicle when ordering.

A

A/C compressor 16, 147-148
A/C compressor install 147-148
A/C refrigerant 17
Accessory-drive belts 16, 150
Air cleaner 15, 151, 152
Air pump 34, 148
Alternator 18, 21, 141, 148
Alternator bracket 138
Anti-seize compound 110

B

Battery 12, 141
Bearings
    edge-ride 61
    inspect 47
    also see types of
Blowby 3
Bore taper 45, 53-54
Bore-wear measuring
    dial-bore gauge method 53-54
    ring-and-feeler method 54
Boring 4, 5, 7, 56-57
Bottom end 5

C

Cable
    lifting 21-22
    starter 20-21
Camshaft 4
    base circle 10, 64
    bearing journal 64
    inspection 63-64
    installation 97-98
    out-of-round 64
    plug 49, 96, 97
    profile 63
    remove 47
    runout 64
    thrust plate 47, 48, 71-72, 98
Camshaft bearing
    chamfer 94, 95
    install 94-96
    mandrel 50, 95
    remove 50
Camshaft thrust plate
    bolts 97, 98
Carbon deposits 4, 5
Carburetor 143, 144
    choke 7, 144, 145
    heat riser 138
    linkage 144, 145
    remove 39, 40
Casting number 25-27
Change level 24
Clearance volume 32

Clutch 39, 136-137, 138
Clutch linkage
    remove 18, 20
Combustion chamber 4
Compression height 30-32
Compression tester 7-8
Connecting rod 29-30
    bearing 5
    broached 30
    center-to-center length 29
    inspect 66, 69-70
    journal protectors 45-46
    numbering 45
    reconditioning 69-70
    spot faced 30
Coolant loss 8
Cooling lines
    transmission 20-21
Compression ratio 32, 34
Core plug, see Freeze plug
Crankcase ventilation 143-144
Crankshaft
    balance 29
    bearing 99
    bearing caps 102-103
    bearing journal finish 59-61
    bearing-journal radius 61
    end-play 103, 104
    front seal 110
    inspection 58-62
    install 98-103
    nodularity 29
    oil slinger 109
    out-of-round 59-60
    rope seal 99, 100
    runout 61, 62
    stroke 28, 29
    taper 59-61
    throw 29
Crankshaft bearings
    install 100, 103
Crankshaft damper
    install 108-111
    key 108
    remove 42, 43
    repair sleeve 110
    pulley 138
Cylinder block
    cleaning 51-52, 58
    decking 56
    identification 27-32
    notch 56
Cylinder head
    assemble 89-91
    identification 32-38
    install 114-115
    milling 78-79
    remove 40-43
    teardown 75-77

Cylinder pressure 7
Cylinder sleeve 55

D

Distributor 5, 6
    armature 124, 125, 133
    assembly 128-133
    breaker plate 124, 125
    breaker-point adjustment 132
    breaker-point type 122, 124, 130-132
    breakerless 122, 124-125, 132-133
    bushing 124, 127, 128-129
    calibration 133
    cam 125-126, 130-131
    centrifugal advance 122, 124-125, 130
    collar 127, 128-129
    disassembly 124-128
    dual advance 122
    dwell 122-123
    end-play 128-129
    gear 126-127, 130
    Hypalon® sleeve 125, 130
    install 144-146
    magnetic pickup 124, 132-133
    primary wire 124, 130, 132
    remove 39
    shaft 126-127, 128-129
    solid state, See Breakerless
    testing 122-124
    vacuum control valve, see PVS valve
    vacuum diaphragm 123, 124, 132, 133
Deck height 29, 32
Deck-height clearance 32
Detonation 4, 47
Diagnosis 5-10
Dial indicator 8, 61
Dial-bore gauge 53-54
Dieseling 5
Driveshaft
    oil pump 73, 74, 111
Dwell tachometer 6
Dynamometer 4, 155, 156

E

Engine
    balance 32
    ground strap 149
    install 139-153
    mount 19, 21, 39, 138
    plate 137-138
    remove 11-23
    run-in 151-152
Engine block 27-28
Engine-mount bolt 19, 139, 141
Exhaust manifold 39
    install 119, 120
Exhaust pipe 20, 142

F

Fan 14, 15, 149
Fan shroud 15, 151
Filler tube
    transmission 18, 149-150
Firing order 10, 117, 146
Flexplate 39, 40, 137, 138
Flywheel 39, 137
Freeze plugs
    install 90, 96
    remove 41, 77
Front cover
    install 108-111
    remove 42, 44
Fuel filter 144
Fuel mixture 5
Fuel pump 18, 138, 139
Fuel-pump cam 44, 108-109

G

Gasket adhesive 93, 118, 119
Gauge
    runout 86
    small-hole 82
    vacuum 6
Generator
    install 148
    remove 18, 21
Glaze breaking 55-56
Go-No-Go feeler gauge 117, 153
Graph
    cylinder wear vs. operating
        temperature 53
    taper vs. ring-end gap 54
Ground strap 18, 149

H

Harmonic balancer,
    see Crankshaft damper
Head gasket 114-115
    blown 4, 8
    composition-type 79
    shim-type 79
Heater hose 148-149
Hood 12, 152
Hot-lash adjustment 153-154
Hot tank 50, 51, 58
Hydraulic lifter
    adjusting 116-118
    inspection and cleaning 64-66

I

Identification
    decal 25
    tag 24, 25, 39

Ignition
    solid state 6
Ignition coil 144, 145
Intake manifold
    baffle 91-92
    cleaning 91-92
    install 118-119
    remove 40

J

Jack 19-20, 139

K

Knocking, see Detonation

L

Leak-down tester 7-8
Lifter bore, checking 56
Lifter
    install 115
    remove 48-49
    wear 49
    also see types of

M

Magnaflux® 58
Main-bearing cap
    installation 102-103
    removal 46
Mechanical lifters
    adjusting 116-117, 153-154
Molybdenum-disulfide 98

N

Noises, internal engine 4, 5

O

Oil consumption 3
Oil-cooler lines 142
Oil filter 142
Oil-filter adapter
    install 96-97
    remove 50
Oil-gallery plugs
    install 96
    remove 49-50
Oil pan
    install 111-113
    remove 42, 43
Oil pressure sender
    install 138, 139
    remove 18-19

Oil pump
    assembly 73-74
    inspection 73-74
    install 111-112
    remove 42, 44
Oil-pump shaft 110-111, 127
Oilite 62

P

Part numbers 26, 28
PCV system 143, 144
PCV valve 3
Performance 3
Pilot bearing 62
Pinging, see Detonation
Piston 30-32
    compression height 30
    cleaning 67-68
    dish 31
    dome 31
    inspect 65-68
    knurling 55
    nomenclature 65
    skirt 5
Piston and connecting rod
    assembly 70-71
    installation 106-107
    remove 45-46
Piston ring
    compressor 94
    expander 105, 106
    filing 104
    gap 104
    install 105-106
    pip mark 105, 106
    twist 105
Piston rings 3, 5, 7
    chrome 56-57
    moly 54, 56-57
    plain cast iron 56-57
Piston slap 5
Piston-to-bore clearance 55
Plastigage® 94, 100-101
Power-steering pump
    install 148
    remove 16
Preignition 4
Pressure plate, see Clutch
Puller 62
Pushrod 33-34, 115, 116
    guide plate 33
    length 34-35
PVS valve 120, 121

R

Radiator 14-15, 151
Release bearing 137

Ridge 45
Ridge Reamer 45
Ring Expander 65, 66
Rocker arm 32-33
    clips 153
    design 32-33
    install 115-116
    nut 75-76
    pivot 32-33, 75, 76
    pushrod guided 32-33
    rail-type 33-34
    remove 41, 42
Rocker-arm rails
    rail-to-retainer clearance 75-76
Rocker-arm studs
    oversize 80-81
    positive stop 35
    screw-in 35, 76, 80-81
Rust proofing 58, 62

S

Sealer 93
Shop manuals 157
Slide hammer 62
Small-hole gauge 81-82
Spark plugs 5-6, 138
    gap 146, 148
Spray-jet tank 50, 51, 58
Starter motor 21, 140-141
Swept volume 32

T

Tables
    camshaft bearing journal diameter 64
    camshaft lobe lift 9
    crankshaft casting numbers 28
    cylinder-block casting numbers 28
    cylinder-head casting numbers 28
    cylinder head changes 36-37
    deck height & bottom-end
        compression components 31
    engine identification code 24
    head-bolt torque specifications 115
    intake manifold torque specifications
        119
    intake manifold vs. cylinder head
        milling requirements 78
    pushrod lengths 35
    taper vs. ring end gap 54
    timing chain and sprocket applications
        72
    valve adjusting sequence 117
    valve spring specifications 89
    vehicle identification codes 26
Tachometer 6
Taper plug 81

Tappet, see Lifter
TDC
    checking for 8, 144, 145
Test
    compression 7
    leak-down 7-8
    power balance 6-7
    vacuum 5-6
Timing chain 70-73
Timing chain and sprocket
    install 108-109
    remove 43-44
    roller type 71, 73
    wear 44
Timing-chain cover, see Front Cover
Timing gear, see Timing Sprocket
Timing light 5
Timing sprocket 44-45, 70-73
Thermactor, see Air pump 34
Thermostat 120, 121
Thermostat bypass hose 147-148
Throttle linkage
    remove 15
Throttle plate 7
Throwout bearing, see Release bearing
Thrust plate, see Camshaft thrust plate
    71-72
Torque converter 21, 23, 134-135
Torque wrench 94
Transmission
    filler tube 149
    front seal 134-136
Tube-nut wrench 14
Tuneup 4, 155-157

V

Vacuum hoses 149
Valve
    adjust 116, 117, 153-154
    burned 4, 8
    grinding 85
    inspection 83-85
    install 91
    lapping in 86-87
    margin 85
    reconditioning 85, 86, 87
Valve cover 41, 119
Valve guide 3, 81-83
    clearance 82
    inspect 81-82
    knurling 82
Valve-guide insert
    bronze 83
    cast iron 83
    screw-in 82, 83
    thin-wall bronze 82, 83-84
Valve lifter, see Lifter

Valve lash 5, 116, 117, 153-154
Valve lift, checking 8-10
Valve seat
    angle 85
    grinding 85-87
    runout 86
Valve spring
    compressor 76-77, 91
    free height 87, 89, 91
    inpsection 87-89
    installed height 87-88, 89, 91
    keepers, 76, 91
    load 87-89
    open height 88
    rate 87
    retainer 37, 76, 91
    shim 88-91
    solid height 88, 89
    squareness 87
Valve stem caps 33, 76-77

W

Water pump
    remove 39, 40
    install 112
    pulley 149
Windage tray 107, 108
Wrist pin 5, 67, 70
Wires
    ignition 6, 7
Wiring harness 144, 145

9-75432117101358